D0829640

Using Wild and Wayside Plants

Using Wild and Wayside Plants

Nelson Coon

Dover Publications, Inc.
New York

Published in Canada by General Publishing Company, Ltd., 30 Lesmill Road, Don Mills, Toronto, Ontario.
Published in the United Kingdom by Constable and Company, Ltd., 10 Orange Street, London WC2H 7EG.

This Dover edition, first published in 1980, is a corrected republication of the work originally published by Hearthside Press, Inc., in 1957 under the title *Using Wayside Plants.* The ten photos have been deleted in the present edition, causing a change in the pagination. The bibliography has been greatly expanded and updated.

International Standard Book Number:
0-486-23936-5
Library of Congress Catalog Card Number:
79-55236

Manufactured in the United States of America
Dover Publications, Inc.
180 Varick Street
New York, N.Y. 10014

PREFACE TO THE 1969 REVISED EDITION

Twelve years ago, when this book was written, it was planned as a guide to the useful plants of the general region of the northeastern United States, but its use by nearly 50,000 readers from coast to coast has proved that the plants which were selected are common to all regions except for a few small areas in our Union. A check of a current book on the edible plants of the Rocky Mountain area reveals that it includes comments on half of the plants discussed in this work. Many of the lesser known edible plants are those found in the dessert areas of the South or in the higher elevations of the north.

Thus, in again presenting this work to my public, I feel confident that it will be found useful by a wide range of people. The increase in the enjoyment of our parks and trails by campers and nature lovers, and the need for developing in city-bred children an appreciation of the bounties of nature, have encouraged my publisher (and good friend), Nedda Anders, to offer this extended edition to a new and growing public.

I have thought that it might be valuable for readers to know more about the extensive list of poisonous plants that should be avoided, and the various uses of wild plants. These topics are extensively covered in the additional chapters especially prepared for this edition. I am sure these chapters will be found useful by the growing millions who have discovered the joys of going camping in the intervening twelve years since the first draft of this book.

I appreciate the numerous comments and suggestions which have been directed to me from time to time in the past decade, and would welcome more of the same from those many readers who are often more knowledgable than I.

NELSON COON

January 1969
Martha's Vineyard
Vineyard Haven, Mass. 02568

PREFACE TO THE [illegible] EDITION

[illegible]

[illegible], 19[illegible]

CONTENTS

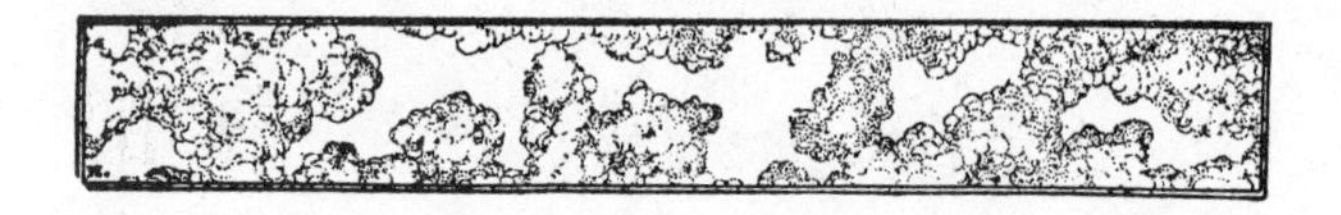

The Bounty of the Wayside

"The earth, that's nature's mother, is her tomb;
What is her burying grave, that is her womb:
And from her womb children of divers kind
We sucking on her natural bosom find;
Many for many virtues excellent
None but for some, and yet all different.
O, mickle is the powerful grace, that lies
In herbs, plants, stones, and their true qualities:
For naught so vile that on the earth doth live,
But to the earth some special good doth give;
Nor aught so good, but, strain'd from that fair use,
Revolts from true birth, stumbling on abuse:
Virtue itself turns vice, being misapplied;
And vice sometimes by action dignified.
Within the infant rind of this small flower
Poison hath residence, and med'cine power:
For this, being smelt, with that part cheers each part;
Being tasted, slays all senses with the heart."

Romeo and Juliet, Act II, Scene 3

Chapter I

SOME GENERAL THOUGHTS ON THE USAGE OF WILD PLANTS

ALTHOUGH THIS BOOK IN ITS ENTIRETY IS (in essence) a *HOW*-TO-DO-IT book, this introductory chapter is intended to answer the general questions of WHO, WHAT, WHERE, WHEN and WHY, the answer to which will not be found in the more intimate discussion of the properties and natures of the various plants.

This book has been written by one who knows through experience the pleasures which can be obtained through an observation and use of the harvest of the highways, and who also knows the truth of the statement made by Walter Beebe Wilder, that "anyone who has been intimate with the hedgerow has had an experience which will last through life."

The answer as to "WHO" can participate in this enjoyment of nature is "Everyone" — Everyone, that is, who has two legs, a bicycle, or an automobile. The use of an automobile, however, predicates speeds much slower than signs will permit and very often excursions into side roads. Increasing traffic in these latter days seems to inhibit the higher speeds and it will, perhaps, soon be much easier to stop by the roadside, as the traffic increases. It is to help those people who want to enjoy nature at the slower speeds, and profit from that enjoyment, that this book has been prepared.

The second of the quint of questions is "WHAT." The basic purpose of this book is to present that answer in detail, but it should be remembered that there are many more useful and edible plants to be found in the region than have been discussed here. The attempt, rather, has been to discuss here those plants which are easier to identify or are more useful than some others. The necessity for such a limited discussion is apparent when it is noted that a recent survey of the naturally growing plants of Massachusetts indicated that there are 2,481 species of plants growing within the boundaries of that state alone. There must, in the very nature of things, be uses for many more of these plants than an arbitrary 100.

Also it is interesting to note that of these 2,400 plants, the estimate is made that about 30% have been introduced from other states or from foreign countries. The fact of the foreign origin of many "native" plants will be noted as one goes through the book and discovers a name such as "Japanese Rose." Such immigrants, including many weeds, have been brought here by intent or more often accidentally.

Beyond the list of plants found in the particular limited area covered by this book, there are, of course, many species which infringe on the area from other sections. A discussion of the useful plants of the southeastern United States, southwestern, or western would include many of the plants of this book, eliminate many, and add others of great interest.

Because it is assumed that this book will be used by the average citizen who is, perhaps, not too familiar with botany, scientific descriptions have been limited to a minimum, reliance rather having been placed largely on the use of line drawings for the purposes of plant identification, added to which illustrations will be the ordinary knowledge that almost anyone would be apt to have, such as the ability to distinguish a dandelion from a maple tree. For further simplification, plants have been grouped according to their general condition of growth; — that is, as trees, shrubs, herbaceous plants and so forth.

Since many people have asked the author, from time to time, for really definite instructions as to how certain plants are to be used, particularized recipes for plant use in food, medicine, etc. have been included in separate chapters, while the more general observations on their characters and properties are included in the individual discussion of a particular plant.

The "WHEN" of the usage of wild plants is naturally a matter dependent on what one is going to use it for. There are certain plants which may be gathered for craft work in midwinter, but generally the period of usage is from spring to late fall, such times being mentioned in appropriate discussions.

Approaching the question of "WHERE," one comes to the problem of the region in which the plants may be found, and for this, recourse has been made to maps. It must be understood that in a map so small, a certain amount of generalization has been necessary, and because each plant has its own natural habitat, there is no use looking on a dry sandy plain for plants normally found in wet places. However, within such environmental limits it should be possible to find the plants (according to the best botanical authorities), in the states as indicated. More specifically, it may be said, as indicated in the title of this discussion, that these are "plants of the roadside." To a great extent, the average country road, (and often the verges of parkways) provides the most happy home for a wide variety of natural vegetation. The fact that on a road running north and south there will be, on opposite sides of the road, both a sunny and shady condition, as well as the probability of a ditch for moisture-loving plants, and often a real stream, means that most conditions for the good growth of a wide variety of plants are present. There are, it is true, some plants which seem to prefer the open woods or the meadows, but in general, one can find a great share of the species discussed here not too far from the highways or byways.

It is good that this is so, because this will mean that plants and their fruits can be taken without the necessity of infringing on the rights of property owners. In the still untouched miles of country roads, there is ample opportunity to secure the wayside bounties without fear of trespassing, but if within a cultivated region, one should always secure permission where there are fruits or other plant products to be gathered.

Such permission is usually granted without question, except where the farmer or owner is dependent upon those same products for his livelihood. Probably nothing but good would come from the use of any plants which are growing on or within the actual rights of way themselves, providing that one does not mutilate trees or unduly disturb the landscape in so doing.

From the "Voice of Experience," — a good way to be sure of collecting such useful plants and their products is to take the advice of Scouting and always BE PREPARED. Standard equipment in any automobile from March until November should be a small market basket or old fishing creel, a suitable knife for cutting, a trowel for digging and, (for use largely in the fall) a pole with a hook for pulling down fruit-laden branches or for obtaining brackets of the edible Pleurotus which are otherwise too high to reach. With this equipment in the car plan on a ten-mile speed when in any likely territory, stop frequently for short prospecting excursions into woods, and keep the eye and mind ever alert to possibilities. With such equipment no Sunday ride in the spring should fail to produce at the least the makings of a "country soup," nor a trip in the summer or fall fail to offer to the housewife plenty of basis for jelly or wine. And it is all free and — Oh, so different!

The answer to the final query "WHY," brings up the old argument as to what is true conservation.

The desirable reasons for collecting and using the products of wild plants are obvious. By this means, the winter larder may be stocked with flavorful and unusual foods in which one will take a great pride, and which in themselves are wonderful "conversation pieces" for the winter dinner table. To this one can add the development of a hobby, the health and pleasures of outdoor exploration, a widening knowledge of nature, and for those with growing children, an unequalled introduction to Nature and natural science.

Against these reasons for gathering the bounty of the wayside, one will hear the cry of a certain type of "conservationist" who will say that this is despoiling the wild. But the more modern and scientific-thinking conservationists are replacing this reasoning by a better one. The wild game enthusiasts, for instance, have *increased* the supply of game, not through "no-hunting" laws but through educated and intelligent self-control. Such self-control should be the approach to using any of Nature's bounty.

In every way the true nature lover will not consider the *use* of wild plants with disfavor, but will have the attitude of the modern conservationist, who understands that true conservation is found in proper use of all natural resources. This point of view was perhaps best expressed in an article in *The Atlantic Monthly* for June, 1952, where Eugene Holden, writing under the title "Our Inexhaustible Resources," has this to say:

"A fallacy is the conception of conservation as non-use. I am convinced that non-use results only in hobbling progress. It will not result in more natural resources for men to use, but less, because it retards the forward march of scientific knowledge. Now it goes without saying that I do not advocate reckless squandering of natural resources. What I do advocate is true conservation—which is not hoarding but efficient and intelligent use."

Another possible reason *for* making use of many wild plants is the fact that in themselves they are apt to contain many elements extremely valuable in the promotion of health. There has been a great deal said in the last few years about the value of organic gardening, and thus under natural conditions it is probable that the food products that one secures from the wayside will have a full measure of all natural elements. More than one scientist has pointed to the fact that diets of certain primitive peoples who use indigenous plants were more balanced than of those who use "store foods" and indeed one survey made several years ago, showed that the children from underprivileged homes in a selected city of Mexico, were more adequately fed from the standpoint of vitamins and essential minerals, than were the children of a selected city in our own country. Another investigator has indicated that from the same standpoint of important chemicals and vitamins, the native population of a country like India is apt to have a better balanced (though often insufficient) diet than those from nations with more "civilized" diets. In both these cases the reason seems to have been that the natives were using food plants indigenous to their country, peppers in Mexico and "curries" in India being high in needed vitamins.

In other categories, too, there is a growing realization that something has been lost in this age of chemistry. The soft pastel shades of the dyes made from wild plants is more and more appreciated, while the medical profession itself is more and more engaged in assessing the successes of "medicine men" of more primitive peoples.

After advancing these arguments the thought suddenly came to the writer: *Do you practice what you preach? Are the products of these trees and plants used in your home?*

Then leaving the typewriter to cool, a search was made to see what products of our American wild plants the average home (which I hope mine is) would have.

On the pantry shelves I found the following:

Canned mushrooms
Wild rice
Powder for Gumbo filé
(sassafras base)
Canned blueberries
Wild grape jelly
Beach plum jelly
Choke-cherry jelly
Hickory-smoked bacon
Maple syrup (homemade)
Clover honey
Cranberry sauce
Blackberry jam
Cocktail crackers
(wrapped in sea-weed)

And down in the basement there is:

Dandelion wine and Elderberry wine.

(This is a list, one notes, that mentions only our own wild plants and omits the usual "herbs" of more exotic origin.)

There could be in my home (and have been) many other food items seasonally used. Dandelions, watercress and chicory in the spring for salads, — asparagus, pokeberry, and fern croziers for greens; strawberries, blueberries, wild black-berries in the summer, elderberries for pie; and many others.

Looking then into the medicine cabinet I was surprised at the frequency with which the names of wildings appeared. Witch hazel finds a first place on most shelves. In a healing salve frequently used, the ingredients were given as: "white oak bark, black ash bark, red sumach berries, marshmallow root, and button snake root;" while on a cough syrup bottle appeared these names: "white pine bark, wild cherry bark, spikenard, blood root, and balm of Gilead." To be listed as "miscellany" was pine oil soap for washing floors, while in my gardening kit is a supply of summer-gathered sphagnum moss for seed planting purposes. Outside the window on this winter day glow the red berries of the deciduous holly and not far away are bushes of bay-berry. With this inventory of the author's own day-to-day use of more than thirty of the plants of this book, it can easily be seen how important is a knowledge of a basic list of useful plants. Our debt to the wild plants of the Northeast is truly more than we realize. Exploring the actual and practical use made of the bounties of nature under the pressure of necessity, an extraordinary account of a modern case of dependence on the wild is given in the true story of one of the characters in the book *Three Men* written by Jean Evans (Knopf, 1954), where she tells of the survival methods of her "William Miller:"

Miller looked on the periods spent with other people as "evil times," to be borne only as a means to an end. Only when he had finished a hitch on a farm or in a city, had a pack on his back, a trinket or two in his pockets, and was on his way to a spell of "living with nature" did his "bad feelings" fall away. . . .

As soon as he made camp, if the weather was good, he sliced his bread, threaded it with string, and festooned the trees with it, leaving it until it was hard and dry. "Then there was plenty for me and the birds." . . .

He built his shelters with whatever materials were available. Best of all he liked a stone hut, but he also used tree branches and, if he happened to settle in the vicinity of an abandoned lumber camp, down-timber. Though he tried to use the coldest months of the year for the business of replenishing his supplies, occasionally, caught in the country in winter weather, he found shelter in the mouths of caves or built his shelter over a dugout, scooped out with a tin can or cooking utensil. "People don't know it, but in some places, if you know how to find them, the ground under the snow isn't frozen, even in below zero weather, and you can dig yourself a little place that's as cozy as a nest." . . .

He supplemented the food he bought with the meat of trapped game, mainly rabbits and squirrels. If there was a farm in the vicinity, he went on occasional foraging trips, drinking fresh milk on the spot, and carrying back chickens and fresh fruit and vegetables. He kept razor blades, soap, and a clean shirt for just such excursions: "so if I was seen walking along a road in the early morning, I wouldn't look suspicious." He also supplemented his diet with wild things growing around the countryside. . . .

"Many times in the spring I'd tap the sugar maple and drink the sap. I'd bore a hole in the tree; then I'd punch the soft pith out of an elderberry twig and blow an air hole into it, and drive it into the tree and put a tin can under. The sugar maple isn't as pretty as the swamp or red maple," he said, "but I love that tree best of all in New England because the syrup is so good. I just love sweet things."

He ate walnuts, hickory nuts, and hazelnuts. The bark of certain trees, though not actually edible, was fine to chew while on long hikes and diminished the feeling of hunger. "The sassafras has a wonderful aromatic taste," he said. "It makes the saliva act up. Once you know that taste, you always look for it. You just crave it. The elm," he continued, "is bitter, but that's an interesting taste, too. Nature is just full of variety. The black birch has a lovely, sweet taste," and nibbling the young shoots of the Southern sourwood tree "is just great for quenching thirst."

He ate wild cherries, plums, crabapples, and berries. "May and June, I had wild strawberries, blackberries, dewberries, red raspberries, and June berries; in July and August, blueberries, thimbleberries—that's the blackcapped raspberry—black elderberries, nannyberries, highbush cranberries." With the help of a Boy Scout book and plant and tree encyclopedias, which he later began to carry, he discovered certain root foods. His leafy vegetables were marsh marigold, watercress, chickweed, pokeweed, and milkweed, squirrel brier, sorrel grass, Russian thistle, and the young leaves of dandelion and clover.

(From "William Miller," in *Three Men*, pp. 117-119)

It much behooves every loyal American then as well as the criminal, for profit and pleasure (and self-education), to discover what is at hand along the waysides, and it is the intent of the pages that follow, to aid in this discovery. It *is* true that many states have taken cognizance of the importance of trees and plants and have designated certain ones as "official state trees," for instance, in Michigan, the apple tree is the state tree; for New Hampshire, the white birch; for Maine, the white pine; for Rhode Island, Vermont, and New York, the maple tree; for Maryland and Connecticut, the white oak; but the real question is, *"Is token appreciation of the importance of trees reaching down into the levels of everyday living?"* If the answer is *No,* it is the belief of this writer that greater use should be made of our wayside bounties, with real appreciation following as a natural consequence.

"All that thy seasons bring, oh nature, is fruit for me.
All things come from thee, subsist in thee, go back to thee."

MARCUS AURELIUS, *Meditations*, Book IV

Chapter II
OBSERVATIONS ON FOOD PLANTS

There are so many plants within the northeastern region of the United States that are valuable for food, that it has been hard to select just those plants which would have maximum value. To a great extent, however, we have included in our selection those plants which seem largely to be most easily available, and which may be had for the taking along the roadside without fear of trespassing. First, and within the limits of this Chapter are given recipes for the use of such plants, because it is felt that food has first consideration in the lives of most of us.

In the case of many individuals (and indeed among whole cultural groups), diet has become so "civilized" that people are often prejudiced against the eating of indigenous plants or "weeds," but anyone who has lived in the vicinity of a large city of the U.S.A. where there is a mixed foreign population, will realize how popular some of the wild plants are to people who are "in the know." The gathering of dandelion greens, of wild mushrooms, of grape leaves and many other bounties of the wild, are so much a matter of national or racial inheritance to many groups that they will go to great lengths to secure these delicacies, even where other food is plentiful. Cannot we, too, as Americans, learn to enjoy these foods?

In learning to identify, pick, and use these various food plants, full attention should be given to identification where such is not a natural knowledge from one's childhood. Certain important poisonous plants are discussed in the last chapter of this work and this section should, perhaps, be read first by the uninitiated before wild food is gathered. Cautions about the use of certain otherwise useful edible plants are discussed in the consideration of the various plant species. In practise this book could best be used by reading about each plant whereever it is mentioned in recipe or formula. A few of the plants used for "greens" contain a bitter principle, which is often removed by boiling in two waters, discarding the first water; but beyond this, little, if any, rectification of natural products is necessary.

The recipes given in the following chapter have been collected from a variety of sources and to a considerable extent have been proven through usage. One's own taste will, however, dictate suitable adjustments in sugar, salt, etc. according to needs or habits. The author would greatly welcome correspondence aimed at correcting, augmenting and improving this present collection of recipes.

For those who in one way or another become interested in this subject and wish to know about other food plants, attention is directed to the more encyclopedic work on this subject by Fernald and Kinsey which is listed along with other works in the bibliography.

RECIPES

Soups

SORREL SOUP

Wash well a quantity of sorrel and put in a saucepan with a little water but do not cover. Cook slowly for about one half hour. Take now four cups of milk to which small pieces of white onion have been added. Cook this in a double boiler, adding two teaspoonfuls of butter, two tablespoons of flour thoroughly blended to avoid lumps. Let this mixture stand after it has been cooked for a while and then add the sorrel mixture; strain, season to taste.

The quantity of sorrel or other fresh greens to use in this or similiar soups depends upon the taste. Various recipes noted in our research recommend "a small bunch," "a half handful," and a "good bunch." Trial and error and your own taste will determine what you will eventually call "a quantity." The above recipe may be used for soups made from similiar wild plants such as dandelion, nettles, etc.

COUNTRY SOUP

Chop one onion and a pound of potatoes unpeeled, and fry lightly in butter. Stir in one ounce of flour, add then a handful of spinach, sorrel, and watercress, and cover liberally with water or milk. Simmer this mixture until the potatoes are quite soft and rub through a sieve; heat the mixture again, add salt and pepper and serve with fresh chopped herbs scattered on the top of the soup.

GUMBO

This is not a recipe for soup, but to call attention to the fact that the dried leaves and young tender stems of sassafras have been used a great deal in the South as an addition to soup. The sassafras is highly mucilaginous and a quantity of the powdered leaves can be added to soup from a salt shaker to provide substantive and wholesome addition. It is this powder which forms the famous "*gumbo filé*" of the South.

POTAGE FIN DE SIECLE

Take a quantity of fresh watercress, sorrel, lettuce, and other greens, and a few new potatoes, all chopped very fine, put over a quick fire in a large pan. Immediately pour over this three quarts of chicken consomme, let boil for twenty minutes and strain twice through a fine sieve. Put this mixture back into a casserole and allow it to heat; season to taste, and add the yolks of four eggs, a gill of cream, and for garnishing add a quantity of hazel nuts and fresh watercress.

WATERCRESS SOUP

Wash carefully one pound of watercress and cook about ten to twelve minutes in boiling water enough to cover. Drain off most of the water, add three tablespoons of butter and cook very slowly for about fifteen minutes longer. In another saucepan, melt two tablespoons of butter and blend in two tablespoons of flour, stirring constantly. Add salt to taste, and peppercorns, and cook five to seven minutes or until all starchy taste has disappeared.

Add thickened juice to stewed watercress and heat for about one minute. Put soup through a fine strainer; return to sauce pan and reheat. Garnish with watercress and serve from an attractive Pyrex saucepan.

NETTLE SOUP

Prepare 2 quarts of white soup stock from veal, chicken or other material. When prepared, add about a pound of young nettle tops and a pound of sorrel tops, which have previously been blanched in a frying pan with a little butter. Small sausages, previously fried, can be added in small lengths, and a little sour cream stirred in just before serving.

PUREE OF SORREL

Pick over and wash thoroughly 2½ to 3 pounds of sorrel, and cook like spinach in boiling salted water. When quite tender, remove from the saucepan, drain thoroughly and rub through a sieve. Put this puree into a saucepan with 2 tablespoons of butter and stir well, simmering for about 10 minutes and seasoning with salt and pepper. Then add more butter, a few tablespoons of cream, stirring and mixing well. This should be very creamy and perfectly smooth when served.

Mushroom Cookery

MUSHROOM SOUP

Pour boiling water over two-thirds cup of dried mushrooms and let them stand for a few minutes, then boil in two quarts of either beef or vegetable soup stock, until tender; remove when done, chop the mushrooms further and return to the stock. Melt four tablespoons of butter in a small saucepan, stir in five tablespoons of potato flour, then, without browning, moisten with a few tablespoons of hot stock, and stir into the soup. Just before serving add six tablespoons of sour cream. This is a Polish recipe.

WATERCRESS PUFF BALLS INKCAPS FIELD MUSHROOMS

DRIED MUSHROOM SOUP

This soup is to be made with mushrooms that have been gathered in the summer and dried. To soften them up, simmer one quarter of a pound or less of the ground mushrooms in a quart of chicken broth for a half an hour. S train this mixture and add one cup of sherry wine before serving. Boullion cubes may be used in place of chicken broth.

FRIED MUSHROOMS

In a frying pan put one tablespoon of olive oil, and, when heated, add a finely minced onion. Let this brown slightly then add a little garlic and other fine herbs as desired. Let these brown for three minutes.

PUFFBALL STEAKS

The large round white fungi that are known as puffballs are a fine substitute for meat as far as taste is concerned if nicely fried. Slice them about three-eighths of an inch thick, dip in crumbs with egg, and fry in bacon fat or butter. Only absolutely firm and pure white puffballs should be used.

MORELS

The morels are quite often found in areas where there has been a brush fire and if a sufficient quantity can be secured they may be cooked by baking or frying in butter. The rich juice when boiled down and thickened is served as a sauce.

INK CAPS

In places where trees have died during the past several years, one will often find in the spring or fall great masses of ink caps newly come up from the dead roots of the tree. If these are selected before they turn black they are a most delicious form of mushroom. They can be cut or picked by pulling off the caps without getting the dirty stem bases. Wash well and cook in an iron spider or saucepan with a little butter and seasoning. They should only be allowed to simmer for a few minutes. The black watery juice which remains may be thickened with flour to make a rich creamy sauce. Some people like to bake these mushrooms in the oven with dry bread crumbs. Stir well and fry for about five minutes. Serve the mushrooms on slices of French toast.

PICKLED MUSHROOMS

Wash carefully one pound of small mushrooms, dry thoroughly and peel them, removing the stalk. Boil in salted water until tender, and drain in a sieve. When they are cooled, stack them in a jar and cover with vinegar, which should have been prepared for ten minutes with about twelve peppercorns, a tablespoon of salt, and a few cloves. Allow to stand until cold. Lastly, add two tablespoons of salad oil. Cover the jar with paper or a tight-fitting lid and let stand in a cold place for a few days, when they will be ready to use.

DRYING MUSHROOMS

Where an abundance of mushrooms are discovered they may be kept for future use by drying them in the sun, removing stems from the very large ones. When they seem thoroughly dried they can be placed in a very slow oven for an hour to complete the process and then packed into clean jars tightly sealed. This is excellent material for winter soups and gravies. They may be made ready for use by soaking in warm water for several hours.

Pickles and Relishes

PICKLED MUSHROOMS

Wash the mushrooms and boil in salted water, skimming them as they boil; then put them in cold salted water and allow to stand for twenty-four hours. Pour off the water and allow to stand for a week in white wine vinegar. Remove the mushrooms from the pickle and boil, with pepper, cloves, mace, other desired spices. When liquid is cold, add the mushrooms and place in a tight container.

PICKLED ASPARAGUS

Cut off the white ends, scrape the green ones, dry them and lay in a broad pan; throw over them salt and a few cloves and mace. Then cover with white wine vinegar and let the stalks lie for nine days; then put the liquid pickle in a kettle and boil it. Put the asparagus into the pickle and allow it to stand for a short period. Then place over a slow fire until the asparagus turns green, but do not boil until soft. Store length-wise in a tight container.

PICKLED WALNUTS

When walnuts are at their full size and still green and soft so that a pin may be run through them, cut a little hole in the side of each walnut and pick the kernel out. Put them into a kettle of water and let them sit over a low flame for 3/4 of an hour, but don't quite boil them. Put them into a mild salt solution and allow them to stand for nine days. Take flavorings such as garlic, mustard, nutmeg, mace, cloves, pepper, ginger, and if whole spices, beat all together in a mortar, and add to the spices the kernels of your walnuts. Boil up a little vinegar with allspice, and after cooling add the walnuts to the prepared pickle, and store in tight containers until wanted.

SASSAFRAS CONDIMENT

Grate the dried bark of sassafras into boiling sugar. This mixture when cool would be a very strong flavored condiment to eat with meat.

APPLE CHUTNEY

Wild apples may be used for this recipe for a delicious meat condiment. Prepare two quarts of diced apples; 1 cup of chopped onions, 1 cup of seedless raisins, 1 cup of mixed nutmeats. Grind the following items in a meat grinder: 2 pinches of cayenne pepper, 1 red pepper, 1 teaspoon curry powder, 1 clove of garlic, 2 tablespoons red chile powder, 2 teaspoons powdered ginger, 2 ounces crushed mustard seed, 1 tablespoon salt, 1 tablespoon black peppercorns, and 1 pound of brown sugar. All of this may be combined with 1 quart of tarragon vinegar. Mix all of the items together and pour into preserve jars. Allow the jars to stand for an hour in a kettle of boiling water; then the jars may be sealed and stored.

WALNUT CATSUP

One hundred young, tender walnuts. Prick and put into a jar with water to cover, and a cup of salt. Stir twice a day for 2 weeks. Drain the liquor into a kettle. Cover the walnuts with boiling vinegar, mash to a pulp, and put through a colander into the kettle. For every quart of this take 2 ounces each of white pepper and ginger, 1 each of cloves and grated nutmeg, a pinch of cayenne, a small onion minced fine, and a teaspoon of celery seed tied in muslin. Boil all together for 1 hour. Bottle when cold.

ELDERBERRY CHUTNEY

Crush two pounds of cleaned elderberries, add one large onion chopped up, a pint of vinegar, teaspoon of salt, teaspoon of ground ginger, several of sugar, and a liberal quantity of pepper and mixed spices. Boil quickly and simmer until thick, stirring frequently. Put into cans at once and seal.

SALADS

WILTED DANDELION SALAD

After washing and drying the leaves and soaking for a few hours in salt water to remove the bitterness put them all in an open iron spider which contains ham fat. Stir the dandelions briskly until they are barely wilted; salt and add freshly ground pepper, and serve quickly when the leaves are completely wilted.

CRESS DRESSING

Take 1 pint thick French dressing and mix thoroughly with ½ pint catsup, one hard-boiled egg finely chopped, salt and pepper to taste, a few drops A1 sauce and 1/3 cup finely chopped watercress (cleaned and washed), using only the leaves.

WATERCRESS

Our native watercress is one of the most popular of winter and spring salad plants, and is also considered a great appetizer. It is eaten plain with salt (after washing well) or as an accompaniment of meat or fish.

DANDELION

Cut off the roots and the older leaves of dandelions before flowering, using only the tender heart portion. Wash and soak in salted water for a time until the leaves become crisp. Then drain and press dry and put in a salad bowl which has been rubbed with a clove of garlic. Add French dressing; serve with hard-boiled eggs cut in quarters and laid over.

> Note: If one has plenty of dandelion plants on the lawn, they can be prepared for salad use by tying the young leaves together when growth is just starting, and inverting over the plant a large flower pot. This blanches the leaves, and takes away some of the bitterness.

Miscellany

DANDELIONS WITH BACON DRESSING

Wash the tender new leaves of the dandelions, cutting them across in several places. Put into a bowl and season with a pinch of sugar, salt, and onion juice or a little chopped onion.

Cut bacon into small pieces, using a half strip for each person to be served. Brown in frying pan. When crisp, remove bacon and add twice as much vinegar as there is fat. When this boils up, pour it over the dandelions and stir. Use scraps of bacon for garnishing. Slices of hard-cooked eggs may also be used as a garnish. A combination of several wild greens may be used in preparing this dish.

SAUCE VINCENT

Chop together several stalks of watercress, a handful of chives, one small onion, and a handful of chervil. Crush spices in a mortar, then pass through a wire sieve. Mix with this green puree, a very thin mayonnaise sauce and serve on cold salmon or similar dishes.

CANDIED SWEET POTATOES

Cook sweet potatoes until tender but not soft. Peel and slice lengthwise. Arrange in buttered baking dish and cover with maple sugar and dot with butter. Add a little water. Bake until potatoes are glazed. Carrots may be prepared in the same way.

MEAT WITH VINE LEAVES

In most of the countries where grapes are grown, the natives have found the use of the grape leaves to be of great value in roasting meat. This method is especially recommended when cooking game birds such as pheasants, partridges, squabs, and so forth. The prepared birds are first seasoned and then wrapped each in two large grape leaves and tied with string. Put them in an earthenware casserole, closely packed with 1/4 pound of butter, 1 tablespoon flour, and a little more salt and pepper. Cover closely and seal tightly. Put in a moderate oven for thirty or more minutes according to the size of the birds, but do not uncover until ready to serve.

SORREL SAUCE

Work a liberal amount of butter into a puree of sorrel, add pepper, salt, and a little sugar and thicken with brown sauce (roux). This is a good accompaniment to veal, etc.

SORREL TURNOVERS

Strip the sorrel leaves from mid-ribs, chop finely, add brown sugar and a few leaves of mint; fold, without cooking, into pastry as for an apple turnover and bake as required.

COOKING WITH PURSLANE

Purslane may be cooked and seasoned like spinach, the tender young branches being one of the most palatable of pot herbs. The new tips may be clipped and the growing plant will then furnish a continuous supply. If the "fatty" quality of purslane is objectionable, the tops may be chopped and baked with bread crumbs and a beaten egg. One writer suggests that because of the mucilaginous quality in the plant, it may be used in the manner of okra to thicken soup.

RICE WITH WATERCRESS

In salt water, boil tender one cup of rice. Drain, then steam for a few minutes more. Wash and dry thoroughly two bunches of watercress, and break into pieces. Fry in two tablespoons of butter over a slow fire. Season with salt and pepper. Place with rice in a buttered baking dish in alternate layers, having rice on the top. Sprinkle with grated Swiss cheese and top with butter; bake until light brown.

BRAISED CHICORY

Secure about 1½ lbs. of the tender center leaves of the chicory plant, wash, and discard the coarser leaves. Put into a thick saucepan and sprinkle with salt; add four tablespoons of butter, three tablespoons of water, and the juice of half a lemon. Cover closely, bring to a boil and simmer gently in the oven for 1 to 1½ hours. When ready, put on a hot dish and pour over a little of the reduced liquid from the saucepan.

FERN CROZIERS

In the early spring where these ferns are growing, the lengthening stalks are used when very young and not over six inches high. Break off as far down as tender, remove with the hand the woolly coating, wash and bunch like asparagus. They may be boiled in salted water or steamed until tender, then season with salt and pepper and serve with melted butter or cress sauce. The "wildings" may often be found in city markets along the Atlantic coast.

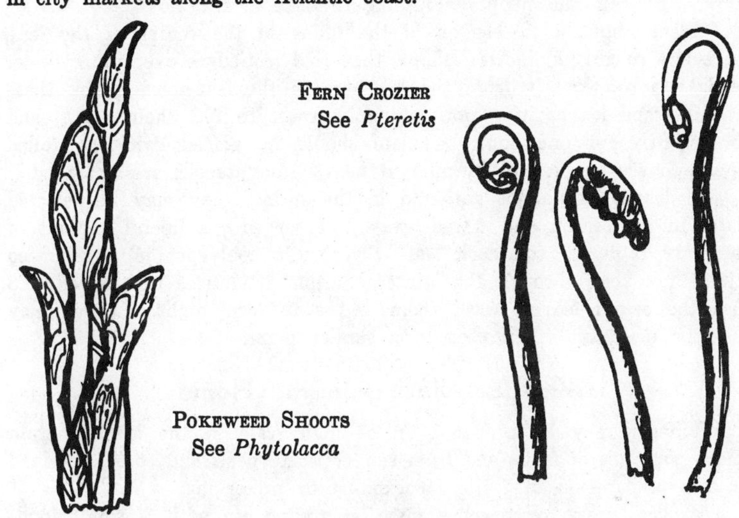

FERN CROZIER
See *Pteretis*

POKEWEED SHOOTS
See *Phytolacca*

POKEWEED

The young shoots of pokeweed should be gathered when about four to six inches long. The developing leaves should be stripped off, the stalks washed, and boiled in two waters. On the second boiling cook until tender, and serve with butter and seasoning, in the manner of asparagus.

POKEWEED IN WINTER

A large root of pokeweed will supply a good crop of sprouts throughout the winter. The procedure is to dig the plant just before hard frost sets in, and plant a half dozen or so roots in a deep box of earth or coal ashes in the cellar. Growth will soon begin and for a period of six weeks a weekly mess of stalks will be provided. (The same procedure can very well be followed with rhubarb, the rhubarb thus "forced" being very tender and pink). It should be remembered in growing these roots in the cellar that they must constantly be kept wet, but the whole procedure may be carried out in a cellar almost devoid of light.

ROLLED GRAPE LEAVES

Grape leaves, to be at their best, should be gathered in June at minimum size, and while still tender. Each leaf can be rolled around a mixture of ground meat and rice, into cigar-like shapes with the ends tucked in. This forms a deliciously tart main course. Here's how.

For a filling use 1 pound of minced meat, (either lamb or beef, with a little suet) to 1 cupful of long-grained rice. Mix this all in a bowl with salt and pepper according to taste.

Place about a tablespoon of the filling in the center of the leaf, give it a roughly cylindrical shape, then fold leaf-edges over the cylinder ends and roll loosely from the leaf-base to the leaf-apex. Stack these rolled grape leaves in a pot, add cold water to 2/3 their depth, and boil gently for one hour. A plate should be placed over the rolled grape leaves to prevent burning, if heavy iron utensils are not used.

If leaves cannot be gathered in the spring, they may be selected later in the season and stored away in a jar with a liberal amount of ordinary table salt on each leaf. Place in a cool spot in a cellar so that they keep through the winter. Salting tenderizes the leaves. To use the salted leaves, wash them and soak over night. Leaves may also be dried and then used in a similar manner.

Vinegar Making in the Home

Vinegar may be prepared on a small scale in the kitchen from many varieties of fruits and from sundry waste peels and cores, obtained in the preparation of fruits for canning or preserving.

Apples, pears, or peaches may be ground up with a small food-grinder, while berries, grapes and other soft fruits may be crushed by hand or with a potato masher. Peelings may be mixed with equal parts of water and boiled until soft.

Press the crushed fruit or the boiled peels through a double thickness of cheese cloth to separate juice from peelings; then add 1/4 pound of sugar per quart. Sugar should not be added to the juices of other ripe fruit.

To each quart of cooled juice, add 1/4 cake of yeast, which should be well broken up and thoroughly mixed with the juice. The juice must not be over 90° Fahrenheit when the yeast is added. Allow it to stand in a stone or glass jar with the lid removed and with the jar covered with a cloth until gas formation ceases; or, allow the juice to ferment in a gallon jug or bottle of any suitable size and plugged with cotton or covered with cloth. This fermentation period usually requires two weeks.

When gas formation has ceased, separate the fermented liquid from the sediment. To each quart of this liquid add about ½ pint of good unpasteurized vinegar. Cover the jar or bottle with a cloth to exclude insects, and allow it to stand in a warm place until the vinegar is strong enough to use. Separate it from the "mother of vinegar" and sediment, bottle it, and cork tightly. "Mother vinegar" is the white, rubbery mass of vinegar bacteria that often forms in vinegar.

Never add vinegar to the fresh juice because it interferes with the yeast fermentation and will result in a weak vinegar. The vinegar must not be added until the yeast fermentation is complete. It will then cause the vinegar fermentation to proceed rapidly and will prevent the molding or spoiling of the fermented liquid.

Vinegar is corrosive. Do not use copper, zinc or iron in handling it. Galvanized ware is extremely dangerous to use, for the zinc coating dissolves and makes the juice or vinegar very poisonous.

CLOVER BLOOM VINEGAR

Allow 6 pounds brown sugar to ½ bushel clover bloom. Add 4 quarts molasses and 9 gallons boiling water. Let cool and add 3 pints hop yeast. Proceed as for other vinegars.

SPEARMINT VINEGAR

Gather clean, fresh spearmint, or peppermint, put in a wide-mouthed bottle enough to nearly fill it loosely. Fill with vinegar, cork, and in about 3 weeks pour the vinegar off into another bottle and cork well. Serve with cold meats. Also good with soup and roasts.

Puddings, Pies, and Cakes

APPLE MINCEMEAT

Chop together 1 pound of wild apples, 1 pound suet; add 1 pound of moist sugar and one of chopped seeded raisins. To this add the juice of 4 oranges and 2 lemons with allspice for flavoring. Mix together and bake in prepared pie crust.

BLACK WALNUT CAKES

Cream together 1 cup of butter and 1½ cups of sugar. Add and mix 3 unbeaten eggs. Sift together 3 cups of flour, ½ tsp. soda, 1 tsp. cream of tartar, ½ tsp. salt. Add this to the creamed mixture alternately with 1 cup of milk. Finally, add 1 cup of black walnut meats. Roll this dough to 1/4 inch thickness, cut into small squares and bake at 450° F. until slightly brown. Reduce the temperature to 350 and finish baking.

MAPLE RICE PUDDING

Beat 3 eggs into 1 cup of maple sugar; then add 2½ cups of boiled rice, 2½ cups of milk, ½ teaspoon of salt, ½ teaspoon of nutmeg, and a cup of raisins. This may be baked in the oven as a custard, or cooked in a double boiler.

MAPLE DROP COOKIES

Cream ½ cup of butter, add 1 beaten egg, 3/4 cup of maple syrup and 2 tbs. of milk. Add 1 cup of chopped nuts, ½ cup chopped dates or other dried fruits, ½ cup candied citron. Combine this mixture with 2 cups sifted flour, 2 tsp. baking powder, ½ tsp. salt. Drop by small spoonfuls on a greased cookie sheet and bake at 375° F. for about 10 minutes.

MAPLE ICEBOX COOKIES

Cream 1 cup maple sugar with ½ cup butter and stir in 2 eggs. Sift 1 cup flour and 1 tsp. soda, ½ tsp. salt, and 2 tsp. cream of tartar. Combine, roll, and keep in refrigerator over night. Slice thin and bake on greased cookie sheet for about 11 minutes in a moderate oven.

BLUEBERRY PUDDING

1 cup of butter, 2 of sugar, 1 of sour milk, 4 of flour, 5 eggs, 1 teaspoonful of soda, and 1 quart of berries. Freshly-ground nutmeg helps to bring out the flavor. Beat the butter and sugar to a cream, and add the eggs well beaten; then the sour milk, in which the soda is dissolved, and then the flour, and lastly the berries. Wring out in boiling water a "pudding-cloth," and spread it in a deep dish; then turn the batter in and tie the bag. Drop this into a kettle of boiling water; boil three hours, turning the pudding often. Serve with wine sauce. This pudding may also be steamed, allowing half an hour longer to cook.

DUMPLINGS WITH GREEN CRANBERRY SAUCE

Make a soft dough with 2 cups of flour, 2 teaspoons of baking powder, 1 tablespoon of shortening, and enough milk for proper consistency. Shape into balls and steam.

Put 2 cups of green cranberries in a saucepan, and cover with water so that they float. Cook until skins break. Strain, sweeten to taste with half sugar and half molasses, and serve hot.

IRISH MOSS BLANC-MANGE

Wash well 1 cup of freshly-gathered Irish moss; let it soak in cold water for 1 hour; then tie up a muslin bag, and put in a tin pail with 3 quarts of fresh milk. Set the pail in a kettle of hot water, and boil for thirty minutes, stirring occasionally. Then squeeze the bag against the sides of the pail with a spoon to get out all the gluten; stir in a teaspoon of salt, ½ cup of sugar, and add any desired flavor. Pour into molds and set in the icebox to cool. Serve with sugar and cream.

BLUEBERRY FRITTERS

Sift 3 tablespoons of sugar with 1 cup of flour, 1½ teaspoons of baking powder, and 1/3 cup of milk, and stir until smooth. Fold in 1 cup of blueberries, and drop from a spoon into hot fat. When brown, drain on a paper bag, and dust with powdered sugar before serving.

ELDERBERRY PIE

Cook 5 cups of fully ripe washed elderberries in ½ cup of cold water over a slow fire, first adding the juice and grated rind of 1 lemon. (Some people use mild vinegar in place of lemon). Keep covered.

Now add 1½ cups of sugar and 1 scant rounded tablespoon of cornstarch. Add slowly until all sugar has been added and the mixture thickened.

Pour this into unbaked pie pastry and cover with lattice strips. Bake in a hot oven until crusts are nicely browned.

Breads

ELDERFLOWER WAFFLES

Wash the flat-topped blossom clusters of the common elder. Shake dry, and dip in waffle batter, then fry in deep fat. Elderflower pancakes can be made, also, by adding a handful or so of the florets, stripped from the stems, to a griddle-cake batter.

BLUEBERRY BREAD

This is a suitable recipe for campers. Take 2 cups flour, 2 tsp baking powder, ½ tsp. salt, and sift well. To this add 1 heaping tbsp. shortening and as many blueberries as the dough will hold. Add milk or water to make a stiff dough, adding sugar as desired. Grease a frying pan and place dough in it. Bake the bottom first by holding the pan over a bed of coals and then prop the pan before the coals and bake the top (using a reflector) to a golden brown.

BLUEBERRY MUFFINS

Sift together 2 cups of flour, ½ cup of sugar, ½ tsp. salt, and 4 tsp. baking powder. Stir in slowly 1 cup of milk and 1 beaten egg. Stir only sufficient to moisten flour; then add 1 cup of blueberries and 2 tbs. of melted butter, with freshly ground nutmeg to flavor. Bake in greased muffin tin in oven at 375° for 20 minutes.

CAT-TAIL-POLLEN PANCAKES

These have been called Sunshine Flapjacks. The time to make them is in late May or early June, when the staminate, upper portions of the cat-tail heads are ready to shed their golden pollen. The pollen is shaken into a bowl or on a clean cloth, and substituted for about half the flour required in any pancake batter.

Jams, Jellies, and Preserves

CRANBERRY SAUCE

Boil a quart of berries with half a pound of sugar and a little under a pint of water for about 10 or 15 minutes until the berries start to crack. Take from fire immediately and set aside to cool and harden.

GRAPE JUICE

Remove wild grapes from the stems after discarding under-ripe and decaying fruit. Place in kettle with a little water, about 1 cup to 10 pounds of grapes. Apply a gentle heat until the fruit is soft, when the free-running juice may be strained through 1 layer of cheesecloth. Do not squeeze the fruit as this will make the juice cloudy. For best results, grape juice should be pasteurized in 1 gallon glass bottles, by being held at 165°F. for 40 minutes in a hot water bath, then sealed, and allowed to stand 3 to 6 months before being again sealed in smaller bottles. At this time the juice may be poured or siphoned out of the large container and into small clean bottles. These should be pasteurized again at a temperature of 160° F. for about 20 minutes. Seal with sterilized caps.

GRAPE CONSERVE

Use only sound, ripe grapes. Separate the skins and pulp, and cook the pulp slightly to remove the seeds. Grind or chop the skins until they are fine, and then cook slightly, to soften. Now combine the skins and pulps and add, for every 3 pounds of fruit, 1 pound of sugar, ½ pound finely ground raisins, the meaty part of 2 large oranges, and 1/5 of the ground peel of an orange. Cook this mixture approximately an hour on a slow fire until it is thick. Then stir into the mixture ½ pound of ground pecan meats. Allow to boil for 5 minutes more, remove from fire, pack solidly in small containers and cover with paraffin.

PRESERVED BARBERRIES

Pick from the stems, clean, and sort 1 peck wild barberries. Bring to a boil, 6 quarts of molasses, then add the barberries and boil about 50 minutes. The fruit will then be clear and plump. Skim out the barberries, and put in jars. Pour enough syrup over the barberries to fill the jars. Care should be taken to see that the molasses does not burn.

WILD STRAWBERRY JAM

To a quantity of freshly picked wild strawberries, add 3 quarters by weight of granulated sugar, and boil fruit and sugar together for twenty minutes, and pour into hot sterilized glasses. Seal immediately.

HAWTHORN JELLY

In proportions of ½ pint of water to 1 pound of fruit, simmer a quantity of cleaned "haws" or hawthorn fruit. Then mash the fruit as for apple jelly, and allow a pound of sugar to a pint of liquid, plus a little lemon juice. This will produce a brown-colored jelly resembling guava jelly in flavor.

CRABAPPLE JELLY

Small wild apples may be substituted for crabapples in making jelly by any standard procedure. The addition of wild blackberry juice to wild apple jelly makes a delicious and beautifully colored jelly.

RUM CHERRY JELLY

Perhaps one of the finest jellies that one can make from wild fruit, is that made from rum cherries which will be found ready about the end of August. The juice should be secured by simmering the berries with a very small quantity of water for about half an hour, and then straining the juice as one would for any similar fruit. This may then be prepared with the use of pectin or by using apples as the pectin producer. Both the color and the flavor of this jelly makes it an ideal accompaniment to a meat dinner.

WILD ROSE JAM (1)

During World War II it was discovered that rose hips were one of the finest sources for vitamin C, and both in England and in the Scandinavian countries, the jam made of rose hips was being widely substituted for oranges. In preparing this jam, the hips of any wild rose may be used but those of the Japanese rose (rosa rugosa) are much larger, and generally available, as the rose has escaped cultivation in many parts of the country (especially along the seashore). Preferably, one should wait until there has been a frost before preparing this jam.

To every pound of hips, add ½ pint of water, and boil until the fruit is tender, when the pulp should be passed through a very fine sieve to make a puree. To this puree, sugar may be added in equal parts by weight, and the mixture boiled until it jellies. Use an enameled-like saucepan for boiling.

SWEETBRIER ROSE-HIP JAM (2)

A recommended recipe suggests using the hips, or fruits of sweetbrier, after they have been touched by frost. Remove the stems and tops, and cook the hips until they are tender, allowing 1 cup of water to each pound of fruit. Put the cooked hips through a fine sieve, to remove the seeds. Add 1 pound of sugar to each pound of pulp, and boil until it jellies. Use only enamel or glass utensils.

A whole-fruit jam can be made as follows. Split the hips, and remove the seeds (the point of an apple-corer works well). Allowing 2 cups of sugar and 1 cup of water for each cup of cleaned fruit, make a syrup, boiling it 3 or 4 minutes. Add the hips, and 2 tablespoons of lemon juice for each cup of fruit. Cover and boil for about 20 minutes. If, at the end of this time, the hips are not clear and the syrup thick, cook, uncovered, a few minutes more.

ELDERBERRY JELLY

Elderberries for jelly should first be taken from the stems, washed, crushed, and then cooked gently to release the juice, after which the juice should be prepared as jelly by adding Certo in accordance with instructions.

CANNED BLUEBERRIES

Blueberries, being so common and plentiful, may easily be preserved for winter use by canning. After being thoroughly washed, place them in clean hot jars covered with a boiling syrup, made from 1 cup of sugar to 2 cups of water. The jars are then placed in boiling water, partly sealed for 20 minutes, taken out and sealed. In this connection, it should be noted that blueberries may be kept during the summer season for a period as long as a month by picking them when they are clean and dry, and putting at once into quart fruit jars, and standing in the icebox. Jars should be tightly sealed.

CRANBERRY JELLY

Boil 1 quart of berries with a pint of water for half an hour, strain the mixture through a sieve, boil 5 minutes with a pound of sugar, and set to cool in a mold.

Candies

CANDIED GINGER ROOT (WILD)

Over a period of two or three days, boil the peeled root of wild ginger for about fifteen minutes, and allow to cool between boilings. On the fourth day, boil the roots until they are soft; then boil them in an equal weight of sugar and water for a short period. Allow the roots to cool in the sugar syrup after boiling for fifteen minutes, roll them in granulated sugar when cool, and place in sealed jars to use as wanted.

CANDIED SWEET FLAG ROOT

A hundred years ago, one of the most common homemade confections was the roots of the sweet flag. The roots should be boiled, more or less continuously, for two or three days until they are soft. Then cut them into small pieces and boil for a few minutes in a medium sugar syrup, roll in sugar when cool, and store them in tight jars after they are thoroughly dried.

GLACED NUTS

Any of the wild nuts which one can secure may be glacéd in a very simple manner as follows: Prepare a sugar mixture with two cups of sugar and a cup of boiling water, 1/8 tsp. of cream of tartar, and a dash of salt. This is boiled until it starts to change color, and the pan is plunged quickly into a larger pan of cold water to stop the boiling immediately. This mixture will become as hard as a rock and glassy clear. When it is wanted for use, the mixture is softened by warming it in a double boiler. Any kind of nuts can be dipped in with tweezers, and placed on waxed paper to dry.

HAZELNUT PASTE

One pound of hazelnuts should be finely chopped or ground in a mill. A mixture of four ounces of granulated sugar and four tablespoons of water is boiled along with a small quantity of candied orange peel for not over three minutes. The nuts can then be stirred into this mixture, and the whole set aside to cool. This flavorful mixture can be used for turnovers, mince pies, and so forth.

MAPLE SUGAR

The requirements for preparing maple sugar from the sap of the maple tree are, largely, a good supply of heat and considerable patience under home conditions. It takes large quantities of maple sap to produce a small lot of syrup, (30-1) but the results are always worth while. No recipe is needed for this process. It is simply a matter of boiling until the syrup is of the desired color and consistency.

In standard recipe books and advertising booklets obtainable in Vermont there are literally hundreds of ideas for utilizing maple syrup. No attempt will be made here to reproduce more than a sample of these many ideas.

One of the nicest experiences in the preparing of maple sugar is to eat some which has been dropped on snow. Of course if the snow can be found while the sap is running, that is the time to do it. If not, the experiment can be tried in the winter with previously made syrup. Some experimentation will be required for this experience in that the syrup should be boiled until, when dropped on snow, it remains on the surface and becomes waxy. When a small quantity has been prepared on a small pan of snow it can be transferred to a block of ice to be used as wanted for the party.

Beyond the above and the obvious ordinary uses for maple syrup, the following sample candy recipes are given:

MAPLE SAUCE FOR PUDDING

Bring to a boil one cup of maple sugar to which a tablespoon of flour, two thirds of a cup of hot water, and a tablespoon of butter have been added. After it has boiled pour it over a well-beaten egg stirring all while adding. This makes a delicious pudding sauce.

MAPLE POPCORN

Take a tablespoon full of butter, three of water, and one cupful of maple sugar; boil until it is ready to candy and then add three quarts of nicely popped corn. Stir briskly until the mixture is evenly distributed over the corn. Keep up the stirring until it cools, when each kernel will be separately coated. Close attention is necessary to ensure success with this kind of candy.

MAPLE "MONEY" CANDY

Boil equal amounts of maple syrup and dark corn syrup until the mixture forms a ball when dropped into cold water. Drop on a buttered pan to form thin discs about the size of a quarter.

MAPLE BUTTERNUT FUDGE

Combine two cups of maple sugar with one half cup of cream. Boil until it strings from a spoon. Mix in one cup of chopped butternuts and cool in a buttered pan.

Drinks

"SUMACH LEMONADE"

Gather a quantity of the bright red heads of the Scarlet Sumac in midsummer and prepare them as follows. Bruise the fruit heads in water with a potato-masher; simmer, but do not boil, and then strain the colored water through cloth to remove the small hairs. Sugar and ice should be added to taste. This drink will resemble a pink lemonade, and provide a slightly acid, flavorful drink. It is said that the various tribes of Indians stored the dry berries to supply an acid drink in winter.

SASSAFRAS TEA

Brew the roots of sassafras trees to produce a deep red liquid. Serve with cream and sugar as desired and you will have what is, to some people, a decidedly palatable drink.

CHECKERBERRY TEA

The fresh or dried leaves of checkerberry (wintergreen) should be steamed in hot water until sufficient flavor is extracted. This tea, when honey has been added for sweetening, is quite soothing for a cold or a sore throat, and withal has a pleasant taste.

DANDELION COFFEE

The dandelion, being closely related to Chicory, can be used to adulterate coffee or used alone to make a somewhat bitter but palatable drink. Dig the roots and dry them, grind and roast them.

ELDERBERRY WINE

Elderberry wine has been made for many centuries both in the United States and in Europe. It is generally considered to be a rather difficult wine to make in a satisfactory manner, as it has a tendency to turn to vinegar or otherwise spoil. With success, however, one has a very beautifully colored and a very delicious wine. A number of recipes are extant for making this wine. Here is one.

Gather ripe elderberries on a dry day, clean from stalks and put them in an earthenware jar or enameled pan. For three gallons of berries thus cleaned, pour on two gallons of boiling water. Press the berries into the water, cover them closely, and leave until the next day, when the juice can be strained from the fruit through a sieve, and then squeeze from the berries any remaining juice.

Measure the juice, and to every gallon add three pounds of sugar, six cloves, and one tablespoon of ground ginger. Boil this whole mixture for twenty minutes, removing scum as it rises. When cool, put into a well-washed and dry cask. Fill the cask entirely and pour very gently into the bunghole a large teaspoon of new yeast previously mixed with a very small quantity of the wine. Have at least a quart of extra juice in reserve to fill up the cask as the wine evaporates. (It may be bottled after about six months of ripening in the cask.) This wine improves with age and is excellent served hot, as mulled wine.

MAPLE SPRUCE BEER

One of the earliest writers on useful plants of the U.S.A., Michaux, gives the following direction for Maple Beer: "Run four gallons of boiling water, pour one quart of maple syrup; add a little yeast or leaven to excite fermentation, and a spoonful of the essence of spruce; a very pleasant and salutary drink is thus obtained." It is presumed from this quotation that one would allow this to ferment and that it be handled in about the same manner as home-made root beer.

CHOKECHERRY BRANDY

In addition to wines which can be prepared from wild fruit (as above), it is possible to prepare a flavorful brandy by adding to a quart of commercial brandy a pint of chokecherry (rumcherry) juice prepared as follows: Strip the wild chokecherries from the stems, crush and boil with a cup of water to a quart of juice, for at least half an hour. This juice should then be strained through a jelly bag and mixed with a quart of sugar to a pint of juice. Then add it to the brandy. This produces a drink with a very rich, yet spicy flavor.

BLACKBERRY BRANDY

Blackberry brandy, although perhaps more desirably known as a medicine, can be prepared in the same manner as chokecherry brandy.

A variation of the above can be provided by adding cinnamon sticks and nutmeg. It should be well filtered through filter paper before adding to the brandy to remove all sedimentary matter and coarse spices.

NATIVE HERB WINE

Two pint nettle leaves
One pint dandelion leaves
One pint burdock leaves
The rind of three lemons
The peels of seven tangerines
The peels of five oranges
One yeast cake
One and a half to two lbs. brown sugar
One and a half gallons warm water

Heat this mixture in a large kettle and simmer for one hour. Stir well and let stand ten to fourteen days. The wine should now be fermented and may be stirred, strained, and bottled. If vinegar is desired, the mixture should be strained and allowed to stand for another seven to ten days. Upon being "soured" (see vinegar recipe), it will be found to be a mild and interesting vinegar for table use.

ELDERBLOW WINE

A beautiful pale yellow wine of delicate flavor can be made from the blossoms of the common elderberry. To a quart of stemmed elderflowers add three gallons of water, five pounds of sugar, and ferment this mixture with yeast cake. After standing nine days, strain and add three pounds of raisins. This should be stored in a cool dark place in an oak cask for six months, then carefully decanted by siphoning and bottled.

DANDELION WINE I

Pour one quart of boiling water over one quart of fresh, clean dandelion flowers and allow to stand for three days. The flowers can then be strained out and to the mixture add the peel and juice of two oranges, one lemon, and a half pound of chopped raisins. Boil all for twenty minutes. When it is partly cooled, stir in one pound of sugar and one half yeast cake. Stand for three days in a warm place, then strain, and allow to stand for two or three weeks, or until fermentation stops. Keep covered during this period, and then decant and bottle. This wine improves with age and should be kept at least a year before using.

DANDELION WINE II

Three pints of dandelion flowers to four pints of water. Pour the water over the flowers. Let this stand for six days. Strain through muslin. Then boil the liquor for half an hour, with two pounds of sugar, one lemon, one sweet orange and one bitter one. Put in a cask or half gallon bottles, for six months, then bottle off.

Chapter III

SOMETHING FOR THE CHILDREN

WHILE FATHER IS BUSY gathering wayside plants for some of the uses suggested in this chapter, while Mother is busy in the kitchen preparing gathered wildings for dinner, what has become of the children? Is there something for them in "plants as they grow" rather than just an admonition to "don't pick the flowers, Dear"? Indeed, and as many country children have discovered for themselves, there *are* many things that the little tots can do with sticks and grass and flowers which, like they themselves, literally grow up over night. In that interim period between the awareness of the great world around one and the days of boy and girl scouting, when the world of nature is first explored, there is that period of phantasy-land when living things may not always be to the child just what they seem to their more realistic-minded parents.

It would be presuming on the prerogative of the childish imagination to suggest here the many magic transformations that plants may make in the world of make believe; and yet, because so much of this transposing takes place so universally, and has done so over many generations, one can dare to suggest a few of the more obvious ideas with which the imagination of the child may be guided.

Take dolls, for instance. They may be contrived of very simple materials and into them may be breathed the breath of life of girlish fancy. The simplest dolls might be little bunches of pine needles fastened at the "waist" and cut off to an even length. If small, they may be stood on a polished surface and be blown with one's breath across the "lake"—like little dancers.

A larger doll can be contrived from a handful of rushes with the top of the bunch folded over for a head and tied, another tie at the waist, and a flower head stem thrust through the bunch for arms. If a dress is insisted upon, the broad leaves of trees can be folded over for a cape, with white birch-bark collars, and tiny hands pinned on, selected from the smallest leaves of the sassafras tree.

If curls for this or other dolls are needed, they can be made from the stems of dandelions. Split the stems lengthwise into several pieces part way up the stem. Moisten by pulling through the mouth and a nice curl will result. Enough stems for a head of curls can be made if you can stand the acrid taste of the dandelion that long.

On a much simpler and temporary level, a doll of sorts can be fashioned—or almost any other shape for that matter—from the omnipresent burdock burrs, which stick to themselves just as easily as to us. They'll stick together to make shapes of any imaginable kind.

As to clothing for more realistic dolls, the little lady can learn to do her own dyeing, just as mother might learn from the chapter in this book on the subject without, of course, fussing over the more technical processes.

The juices of many plants make dye quite sufficient for this purpose; Elderberry for purple, Spiderworts for blue, Peony juice for red, Live-for-ever for green, etc.

What about a home for Dolly? Children can, as they have for millenniums, figure this out for themselves. The house may be a cleared space between the roots of a big tree; the table, the bed, and chairs the nearest flat rocks; and convenient twigs, the tools of living. But have you ever thought of suggesting the nice big wooly leaves of mullein as blankets for the beds, or pillows made of the silk of milkweed pods (in fall) or the catkins of willow trees (in spring)? The empty milkweed pods are wonderful cradles for the baby dolls, while a half of the same pod makes a delightful fairyboat — even better and simpler than one made of birchbark.

Around the dollhouse a practical broom may be made by tying needles onto a stick; a dustpan can be a piece of birch bark. Baskets for the dolls may be made of nutshell halves, while a useful teapot can be made of large acorns with sticks inserted for handles. The teacups themselves may be contrived of rose-hips with bent pins for handles. The possibilities are, in short, limited only by the children's imagination and yours — if you have more than they.

Finally, for "play" food for the dolls' meals there is plenty for all in any summer meadow. Buttercup petals for butter, the seedheads of members of the mallow family for "cheese", dried dock seed for ground coffee, seed stalks of plantain for Lilliputian sweet-corn, and pieces of the head of red sumach for well-done croquettes. The centers of members of the daisy family make good-looking pumpkin pies, with berries in variety serving as simulated dips of the favorite ice cream.

For the little folks themselves at tea, the "sumach

lemonade" recommended under *Drinks* in Chapter two is ideal and different. For sweets on a summer day, nothing is so much fun as sucking the sweetness from the bottom tips of the florets of clover, or, in the spring, from the honey-laden tips of wild columbine. In the fall this same delight may be found in the fragrant flowers of the Japanese vining honeysuckle. Shoots of brier, of blackberry, and of birch are a good nibble in the spring, while the summer gives us wintergreen leaves and the bottom ends of timothy grass, as we walk through the field.

It may perhaps be noticed in reading the last paragraphs that mention has been made of some plants outside the strict list of this book, but they are such common plants that the lapse may be pardoned. But, going from the exclusively wild plants into the flower garden, it is hard to resist mention of the very lovely tiny dolls which may be made by just a little fixing of upside down hollyhock flowers; they provide not only the dolls themselves, but dolls all dressed up in their satiny best. A little sewing up at the "waist" will make these dolls and if our dressmaker is a simon-pure outdoor person, she will find the thread to do the sewing by pulling off a few very stout silk-like threads from the leaves of the common yucca, which few gardens are without.

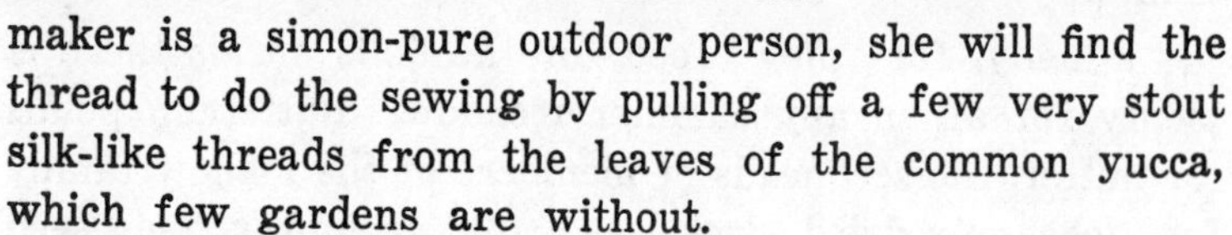

Turning now from the dolls, we come to the old problem of childhood — *'tis and 'taint — are and ain't* — questions which must be decided by the Gods. Here again we can call on plants.

With any ray-petalled flower, questions are easily decided by the old *Yes* — *No* plucking of petals until the remaining one gives the answer. Quicker even than that is the fight of "roosters", the locked heads of violets being pulled until the one decapitates the other. Three or four lengths of grass, held in the hand with the tops even, invite selection of the longer one to make any decision.

More individual matters can easily be decided by other flowers. Do you like butter? — *Yes,* if your chin reflects yellow to a buttercup held up to it. How many children will you have? A perfect dandelion-seed ball blown at three times will, with its remaining seeds, give the correct answer. The same question is decided by squeezing out the florets of the daisy on the back of the hand, throwing them up in the air, and catching all you can. The answer may imply more children coming than one could care for — but what matters at this age? And always one hopes to find the four-leafed clover that is the sure sign of perpetual good luck. But just in case you feel too self-assured about it, be certain to knock on the nearest piece of wood. And so one could go on.

What has been mentioned so far on this subject is in the realm of the imagination — what about the more practical? Well, as most farm boys know, there is a good substitute for chewing gum to be found in the gum exuded from cherry, plum, and spruce trees. It may not be too sweet or tasty but it does exercise the jaws and no harm done. The girls will use the same gum as an imitation topaz when stuck onto the finger as a ring, or made into an earring.

Jewelry can be made, too, from various plant parts.

A chain or necklace of clover blossoms can be put together by tying a loop in the stem just behind the flower and further working this into a succession of cloverheads. A nice jade necklace can be made by cutting dandelion stems into sections and stringing these into a pliable necklace. Pine needle clusters can also be strung to make a green fringe with a very savage, tropical character.

But what about the group which has outgrown the dolls and is not ready for the many interesting but often mundane uses of our 100 plants? The group of the "scouter" age, perhaps? This is the natural time for the developing interest in Indians, Scouting, Daniel Boone, and all the other frontier men who make a perpetual appeal to youth. Ideas for this group are scattered throughout the chapters, but here in somewhat random order are a few additional ideas.

Signalling: A fine shrill whistle can be made by using a thick, strong blade of grass. This should be stretched tightly between the thumbs and blown upon. Variations in material and blowing will produce a variety of pitches and intensities.

A pocketable whistle can be made of a willow twig selected in the spring when the outer bark is soft and can be rubbed off. After the twig is dry, cut a whistle hole in it just a little from the end, and hollow out an airspace from the hole to one end. It may take some experimentation to make a perfect whistle.

An even simpler whistle can be made from the stem of a wild onion, played like a flute.

Guns: Pop or blow-guns can be made from the stems of elderberry bushes, which are easily hollowed out by pushing the pith out of the stem with a piece of heavy wire or an oaken twig. A gun sound effect can be made by holding a tender leaf or rose petal over the lips and sucking in strongly and quickly.

Torches can be fashioned from the heads of rushes, picked as the heads are turning brown. Make certain that the head is wired onto a stout stick so that when in use the flame does not burn it off the thin stem and cause a bad hand burn or a fire.

MISCELLANEOUS PROJECTS

Place Cards can be made by gluing acorns onto a card, drawing a face with ink, and adding hair of wool and a little cap.

Tops can be made from acorns, by pushing a long, sharp, thin nail or stick through the center, and then spinning the top with the fingers.

Flowers may be preserved for winter decorating by drying them in a box of powdered borax or in a mixture of three parts of fine dry sand and one part of borax. If this is done carefully, laying the flowers with stems and leaves into the box and covering them completely and individually, the flowers and foliage will come out when dried in almost their natural color and shape, beautiful for a winter bouquet.

For *Party Fun* or decorations here are a couple more ideas.

Decorative birds can be made of milkweed pods using sticks for legs and black pins for eyes, and as much imaginative make-up as possible.

Cat-tail foliage can be made into a flotilla of ducks for "racing" across a bathtub pond. Fold the leaf into an ellipse and add a leaf bent over to a point for the head. These ducks will swim nicely.

An interesting artistic souvenir of summer pleasures can be made by writing or drawing on the pure white surface of the bracket fungus that grows on many trees. As the fungus enlarges and hardens, the drawn lines will turn dark and provide a permanent picture or message of the *John Loves Mary* type.

The above suggestions by no means exhaust the possibilities of utilizing the bounties of nature for the pleasure of children. But it is hoped they will show how the young as well as the older folks can find much of interest in the commonest of wayside plants.

"On every thorn delightful wisdom grows;
in every rill a sweet instruction flows."
YOUNG, *Love of Fame.*

Chapter IV

PLANT CRAFTS

ONE OF THE FIRST USES FOR PLANTS after Eve gave Adam an apple to eat was the club which Adam cut himself for defense. Since that time man has done a lot of cutting of trees and shrubs not only for clubs but for an increasing number of uses. Of course, many of these uses have become the starting point of great industries, but of these we will have little to say. Rather shall we point out in this chapter some of the ways in which trees and other plants of the roadside may serve the needs of the tourist, the traveler, the camper, and the homemaker. Here also, as in the other chapters, we shall endeavor to confine our consideration only to those plants which constitute the selected group under discussion in this book.

As will be seen, the variety of items which may be made from easily found materials along our roadsides includes such diverse things as canes and candles, dolls and decorations, utensils, twine, and cleaning equipment, and, with imagination, many other things that the necessity of the moment may suggest. The prime lesson that this chapter will endeavor to suggest is to learn to be at home in the woods with friendly plants and to learn to investigate the possibilities of their use in our lives.

The importance of such knowledge of plant, and woods-lore was emphasized during World War II, when all flyers were given an intensive course and indoctrination in survival in wilderness areas. For today's civilian, the thrill is in learning to appreciate and use the great bounty of nature which is everywhere at hand.

WILD PLANTS IN HOUSEHOLD ECONOMY

The obvious uses of plant material in the home are understandable when we look around us and see the chairs and tables, bowls and baskets that are so necessary in homes throughout the world. It is interesting to note that in today's strictly "modern" American home wood is again being increasingly used not only for construction, but for decoration. The utilization of commercial lumber is obviously beyond consideration here, but, as we go down the list of trees and plants many other interesting and non-structural uses suggest themselves; in household economy and home decoration, for example. The reader may well suggest that in most uses here mentioned, modern industry has provided a better substitute, but for the record and the benefit of those who like to experiment, here are some suggestions.

Bayberry Candles

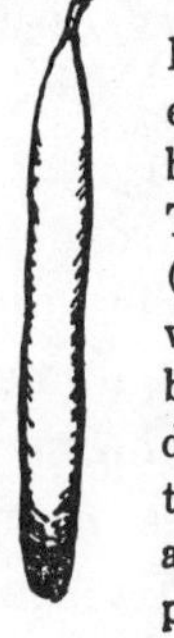

To make candles, the berries of the bayberry are collected when grayish white in color, along in October or early November. The wax may be obtained by boiling the berries in water, and letting the wax harden when cooled. This wax may then be melted and used in its pure state (or mixed with tallow), for candle dipping. The bayberry wax does not produce as bright a light as regular candles, but the perfume which is given off in burning, is very delightful and helps to make for Christmas cheer. Parenthetically, it may be said that the wax of the bayberry may also be used in making soap, substituting in whole or in part for the normal animal oils used in soap making.

Ink

There are several plants which are a source of ink, and in the days of our grand-parents, most ink was so made. One of the present unlikely, but perhaps understandable sources for ink would be elderberry. The berries of the elderbush, or elderberries, certainly stain very easily. One old recipe says that "elderberries stewed with copperas, vinegar and alum, makes an excellent ink." In general, however, the usual source of natural ink is from the tannins which occur in a number of plants, but more especially in sumac and oak trees. Ink from oaks is made available through the gall nuts which are easily found in any woods. This gall nut is an abnormal growth caused by an insect and is usually found as a perfectly round black-powder-filled ball about one and a half inches in diameter. Sumac leaves are another source. Sumac leaves may be gathered in late summer or early fall before they change color, and then dried in the air. The tannin may be extracted by heating the leaves with water in a steam bath (double-boiler) for about eight hours; but do not use boiling water. A little preservative such as thymol should be added to prevent spoilage. The solution so prepared may be made into ink by the following formula:

10 parts of liquid gallic acid.

15 parts of ferrous sulphate solution made at the rate of one part of ferrous sulphate to seven parts of water.

1 part tartaric acid.

3 parts soluble blue color, with additional water to thin it out as needed. In preparing these solutions, always use distilled or rain water.

The cold-water extracted gallic acid from oak galls may be substituted for the sumac solution, and first attemps should be made on a small quantity experimental basis. Ink may be obtained without the use of additional soluble color, but it will make only a faint mark until age turns it dark. Any of these tannin inks are absolutely permanent, as we know from the well preserved writings done with them in colonial times.

Moth Preventives

The common use of thin boards of red cedar as a lining for closets, well indicates one of the most interesting uses of plants. One does not, however, have to have a closet lined with red cedar, to keep the moths away. If it is possible to secure old pieces of red cedar tree-trunks, and cut them up into small pieces and put them in a bag at the bottom of a storage drawer, there will never be any moths, as the cedar odor is very penetrating and long lasting. Crushed dried leaves of the Sweet Gale are also anathema to the moth.

Rush Seating

One of the first uses of wild plants among early settlers in this country, was in the gathering of rushes (Typha latifolia) for rush seating. The antique chairs so seated are much in demand today, and bring high prices. The fact that such rush seating is good after a hundred or more years, shows the permanence of this material.

The rushes are usually gathered in midsummer when the plants are fully mature, and carefully dried in the open air to prevent molding. They are then stored in bundles and then soaked before using to make them pliable. It is not possible here to give the complete instructions for the process of rush seating, but any library would provide books with this information.

While on the subject of this particular plant, it is

also of interest to note that the down from the heads of cattails was much in demand during World War II, for use in life preservers and has also been used in the past for the filling of bed quilts. It is said also, that the Indians used this down as diaper material.

Hickory Smoke

Although it is now possible to buy in the store a preparation of material which in its flavor simulates the smoke of hickory chips, and although it is not likely that the average reader will have the equipment for curing bacon, it is interesting to note that the fine flavor of hams and bacons is secured by using a certain proportion of hickory wood in the smoking of meat. In fact, at way-side stands, the writer has seen bags of hickory shavings sold for the purpose of using on home grills where hickory flavor is desired in outdoor cooking. It will be noted that the word "shavings" is used, because when the wood is cut into small pieces it provides a much greater amount of readily available volatile oils to penetrate the meat.

Nut Oils

Although it is quite possible to buy plenty of cooking oil of plant origin it may be of interest to note that the Indians used the oil from nuts very widely for cooking. The kernels were opened, pounded and boiled in water, the oil then rising to the surface. The oil was skimmed off when cool and used variously for cooking or as butter. It is the volatile oil from the beech nuts which gave the name to the products prepared by the famous Beechnut Company. In fact such oil made from beechnuts, hickory nuts, butternuts and black walnuts, was used by the Indians not only for culinary purposes, but for preparing paints, and as a cure for sunburn.

Cordage

Considering that no synthetic except nylon has been found to replace the plant material used in cordage, one might well say that such a use of plant fibers is of prime importance to our economy. Our cotton cord, our sisal rope, our hemp rope and our silk thread, are all naturally produced materials for which there are but poor substitutes. Among the plants of our region there are a number which may be used for similar purposes on at least a trial or emergency basis.

Among the Indians the inner bark of the basswood tree was a favorite material for making rope and twine. Long vertical strips of bark were soaked in water for a week or ten days, and these strips were then boiled in wood ashes to make them stronger. Then when a certain amount of disintegration had taken place, the fibers could be separated by running a fingernail down through the bark. From this a strong fiber was obtained, which was woven into rope. A good twine from basswood may be made by taking two of the thin strands and rolling them together in one direction between the palm of the hand and the thigh. As the ends of the pieces are reached in twisting, more threads of unequal length are added and rolled gradually, until a long piece is obtained. At uneven places fresh strands are added and finally a strong long piece of twine is produced. Sections of this twine may then be braided together, three strands at a time, to make strong cordage.

The fibers which grow in the swamp milkweed, found at the edges of most wet areas in this section, are extremely tough and may be made into twine and cordage with the same procedure as given for basswood. Another plant, the Indian Hemp, also contains reasonably good fibers for cordage and cloth.

Brooms and Brushes

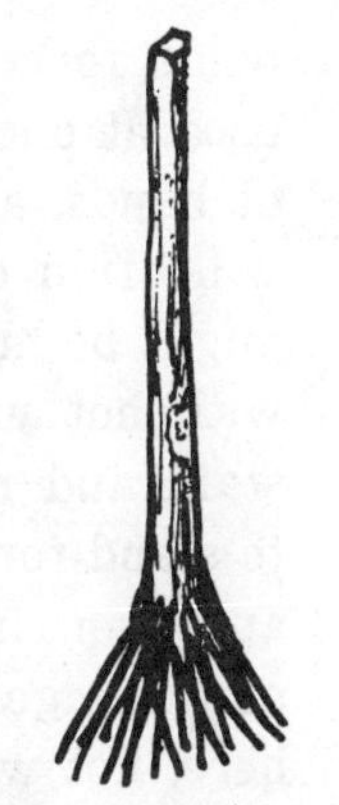

Almost everyone has seen, in museums, replicas of the brooms which were used in colonial times, made obviously, from twigged branches of various trees. It is said that the Iroquois Indians made a broom by taking a large stick of ash, long enough to form both the handle and the broom, and soaking one end of the ash in water for a long period in order to soften the wood. When soft this end was beaten with a maul until it spread it into long pieces, which together formed an efficient and long-lasting broom with a handle as an integral part. For more practical purposes, one can easily make an emergency broom by using the branches of various evergreens or (in the winter) leafless top branches of many shrubs, tied together on the end of a stick.

Scouring Brushes

The Horsetail plant (the scouring rush) contains, as noted elsewhere, a high mineral content, so much so, that it makes a satisfactory scouring brush for cleaning pots and pans. The Indians knew this and would bind a number of the stiff stems together to make a scouring brush, while the stems themselves were often used as we would use sandpaper.

Stain Removal

Because the usage of many wild plants is bound to cause the staining of hands or materials, this chapter would not be complete unless mention were made of the methods of stain removal, and more especially as not all stains should be treated in the same manner. In the case of *fruit stains,* soap should never be used, as this tends to set the stain. Rather, boiling water should be used at once to pour over the stained area, or in the case of hands, a piece of lemon rubbed over the stained hands will often dissolve the stain. If wool is involved (which might be hurt by boiling water), the stain can be sponged with hot water, and then a little glycerine put into the stain and rubbed lightly between the hands. Allow this to stand for a few hours and apply a few drops of vinegar, and then rinse off thoroughly in water. The *stain of grass* and foliage should be treated in a different manner, and here, hot water and soap may be used, rubbing the stain well. If this is not completely successful, a little bleach water or hydrogen peroxide may be used. Benzine, denatured alcohol or cleaning preparations may also be used on delicate materials.

Soap

One of the plants in this book is called Soapwort, and, true to its name, it contains a substance known as saponin. A solution of this material produces a soapy froth which makes the plant available as a soapy substitute. In emergency the plant may be crushed, and the extraced juice used as a substitute for soap. In olden times this juice was frequently added to beer to help make a froth, but it is likely that this is a quality which has small importance in these days.

Saponaria — Soapwort
See Page 198

MISCELLANEOUS CRAFTS

Bird Houses

No book on any aspect of nature could or should ignore our bird neighbors, and of the literature on the subject of bird houses there is no end. These comments are made because the joy of befriending the birds is something no one should miss, and the use of plant materials comes into the picture very strongly, for many birds seem to be most happy when they are amid what must seem to them like "natural" surroundings. Rustic bird houses which look like a log with a hole in it, are very common and attractive to the birds and almost anyone can make such a birdhouse. They do not have to be fancy, nor the edges perfectly even. The general idea is to have a roof, floor, walls, and a small "room" inside for the eggs to which access is given by a "door" of just the right size for each species. The general tendency is to make birdhouses too large, for most birds want "one room only." Bird books provide full information as to sizes of houses and entrances.

Place the birdhouse in what seems to be a safe place, well away from the reaches of the family cat, and if possible, near enough to food and perching sources to be attractive to the birds and to be seen from the home. Many of the wild plants mentioned in this book are food sources for our birds, and may be planted for their enjoyment. Among the fruits of which they are especially fond, (mentioned in this book) are the fruits of the dogwood, the poison ivy, the sumac, the seaside rose, the wild apple, grapes, red cedar and wild cherry.

Among the seeds which birds like, are those from the burdock, the dock, mallow, pigweed and lamb's quarters. Of course, if you especially want to attract many birds, you should additionally provide them with a diet of sunflower seeds and cracked nuts. Persons interested in birds may secure free information on the subject of wild birds and bird-house-making from the National Audubon Society, which has chapters in most cities. Consult your phone directory.

Plant Boxes

One of the most common uses of tree trunks is that of the utilization of the white birch for making plant boxes. Considering the use of birch bark, it should be remembered that completely girdling a tree of birch bark will kill the tree, but it is usually possible to secure permission to cut down whole trees or certain sections of trees from dense growths in birch woods.

Garden Stakes

The author's first contact with the use of wild plants came as a boy, when he was required to go to the woods to secure top growth from almost any wild shrub, which would make suitable material for "pea-brush" and there is still nothing better for the purpose. Garden peas or sweet peas both require staking for best results, and the twiggy branches from the tops of wild shrubs are absolutely ideal. When this is used, laborious stringing is obviated. Again the use of wire mesh will often, in summer sun, burn the tendrils of delicate pea growth. Beyond this use of wild shrubs for brush there is always the need of stakes for tomatoes and other plants, and here again, the undergrowth of almost any woods will furnish desirable material, the size and diameter of which will depend upon the plant for which they are used.

Raising Seedlings with Moss

One of the very useful plants of our book is one of the most primitive and humble ones, the sphagnum moss. The dried moss has been used for years by florists as a base for wreath-making, but it was not until recently that its great value has been demonstrated for horticultural purposes. A Department-of-Agriculture leaflet (No. 243 — 1944) gives detailed instructions for the use of granulated sphagnum moss for seed germination, and in this it is recommended that the fresh-gathered moss be shredded very finely by rubbing through a sieve or otherwise made fine by grinding. This fine moss may be mixed with sand and used in pots or flats as an ideal medium for the sprouting of difficult seeds. The fact that this moss can retain twenty times its own weight in water, means that it can hold water in suspension for seed use over a long period and thus obviate the necessity, sometimes dangerous, of frequent watering. This also means that the seed does not have to be covered so deeply, which is important with a very fine seed. Again it is thought that the natural acid reaction of this moss is a factor in the control of the damping-off disease which frequently attacks young seedlings, and thus its use dispenses with the necessity of soil sterilization. Easily germinated seeds like tomato, petunia and so forth, may be started in a combination of sphagnum, sand and soil, and then transplanted to a richer mixture of the same medium for growing on.

One nurseryman of Massachusetts who specializes in growing many difficult species of the American holly, utilizes pure live sphagnum moss as a growing medium, adding only chemical nutriments as food. He has been able to produce a rapid growth in otherwise difficult plants, by rooting and growing his plants in tin cans (one quart) which have first been painted and had holes punched in the bottom for drainage. A final advantage to the use of this method is that the rooted plants which have been grown in these cans can be planted out of doors with their ball of sphagnum, thus making available moisture for the difficult period of summer dryness which usually follows planting time.

MISCELLANY

Swizzle Stick

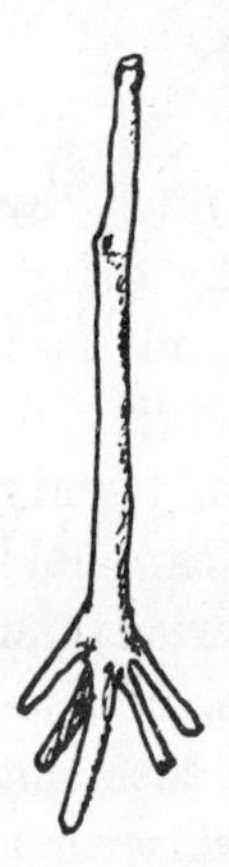

An egg beater can easily be prepared from natural materials by finding in the woods a young oak, maple or other hard-wood tree which is perhaps two or three years old, with a trunk diameter of one-quarter inch. If one digs this up, it will be found to have some branching roots, and the roots thinned evenly to about one inch in length, this stick can be revolved between the hands as an excellent egg beater. Known as a Swizzle Stick, it is used largely in the West Indies and in Mexico, for beating up chocolate and other drinks.

Kindling Fires

Mention need hardly be made here of the possibility of using material from the woods for fires for cooking or heating. Pine cones; of course, are excellent material to use in starting a fire, but care with fire must be great, and all conservation rules carefully followed, for the danger from fires is overwhelmingly a national problem of first importance. All books on camping and scouting also give full descriptions as to the use of evergreen and other boughs for bedmaking in the woods, and a separate chapter in this book is devoted to the medicinal values of many wild plants.

There is however one phase of wildwood wisdom which is perhaps not properly included within this consideration of plants, but because this is a book for wayside travelers, is important enough to find mention here, and that is the matter of the finding and use of drinking water. Almost everyone knows that all water found in the wild should be boiled when used for drinking purposes. In one of the survival pamphlets issued during the war, the government pointed out to the military men that running water or even a spring was not necessarily pure water, because its source is not known.

In practice, however, it may be assumed that water secured by digging a hole and reaching the actual "water table" will be pure, *if* no human habitation is in the vicinity. Water from large lakes exposed to the sunshine also is generally safe to drink if taken at a distance from human habitation. In the woods, a spring issuing from a rock is usually safe unless there is a pasture above. Rain water is pure, and water which is obtained near the seashore by scooping out a hole in the beach at low tide, and allowing the water to seep in, would also be pure, although possibly, somewhat brackish. In digging near the seashore, the best spot for digging is in a low basin where drainage from the land is concentrated.

Fresh water will be found first, when you dig near salt water, as it is lighter than the salt. The other and undersirable alternative is chlorination, for which tablets are obtainable at drug stores. The principal thing to remember is that extreme care must be used in using *surface waters of any kind,* and more especially near habitations, where drainage from sewage and stables provide a ready source for contamination.

A further hint would be that, if the water which is to be used comes from a swamp or other low place, and is smelly, it may be made more palatable if when boiling, small quantities of charcoal are boiled with it. Charcoal is usually available in camp sites or in burned spots in the woods. If no water supply be at hand a second source for the safe satisfaction of thirst comes from the watery sap of many plants. This has been known to native peoples for generations. A wild grapevine, for instance, will produce water at almost any season. The method for securing a drink in this way is as follows:

Reach as high as you can on a big wild grape, and cut a deep notch in the vine or cut it off completely, leaving the severed ends elevated. Then cut the vine close to the ground. This should give you a water tube six or seven feet long, from which water will begin dripping on the lower end. When the water stops, cut off another section at the top, and more water will drain out. Do not cut the bottom of the vine first, because all or part of the water will be lost, as the water will tend to ascend rather than descend. This method can be used in any part of the United States in summer or fall, as wild grapevines of large size are common on our roadsides.

"He that can draw a charm
From rocks, or woods, or weeds, or things that seem
All mute, and does it — is wise."

BRYAN WALLER PROCTER, "A Haunted Stream"

Chapter V

DYEING WITH ROADSIDE PLANTS

ONE HAS ONLY TO SEE WOOL which has been home-dyed and spun in the mountains of North Carolina, or from Scotland; or note the brilliance and fastness of ancient cloth and rugs from both the East and South America, to know that modern chemistry has much to learn from more primitive times. With the modern revival of an interest in the arts and crafts of our ancestors one of the most interesting fields of exploration is the matter of dyeing with native plants. This chapter is made a part of this book not with any feeling that our readers will want to go out and buy raw wool, dye it, and spin and weave their own clothes nor perhaps may ever do any dyeing, but because no survey of plant values can ever ignore this fundamental usefulness of wild plants.

With the general necessity to limit the survey to the hundred plants under discussion, it must be remembered that there are many, many, *other* plants that especially and solely are dye plants, the use of which would greatly enlarge the circle of colors possible. Yet even within the dye-plant list of nearly forty of the selected one hundred, the range of color possibilities is wide and a series of harmonious hues may be obtained which will blend each with the other in a way that no aniline dyes ever could. They generally do not fade nor give unpredictable effects under artifical light and as some one said in a jingle:

"Home-dyed colors kindly meller down,
Better than the new ones fetched from town."

It is the hope, then, of the author that this chapter may stimulate some readers to try their hand at this intriguing hobby of home-dyeing. Enough basic information is provided in the pages that follow to show general procedures and possibilities. Your public library surely has some of the many authoritiative books on this subject, a few of which are listed in our bibliography. All writers are in agreement that the one predictable thing about home dyeing is its unpredictability in both results and in shades of color; and it is that very quality of uncertainty which makes artistic results so likely. Not only are the results variable but so are the recipes given for dyeing; writers are all agreed on this fact and quite likely, as well, it is that variations in color are due in part to the different chemistry of plants, according to location and season. One authority, (Pope), sums it all up in these words:

"Vegetable dyeing is not an exact science. It is instead, an art — a fascinating art, in which each new experiment is beautiful. Commercial dyeing produces no color with the elusive beauty seen in old tapestries."

There is, in dyeing, about four principal processes which one needs to consider: collection of material, preparation of material (mordanting) and dyeing. Some suggestions under each of these processes are herewith appended:

Collection

For the novice there will likely be considerable trial and error in the selection of plant materials for any particular color. Most dye-enthusiasts collect material widely and make small batches of dye as samples to provide a key for future use. This is a practise to be recommended, as the dyeing of small samples will provide the period of experimentation needed in any craft work, which prevents later wastage.

Under the headings of the various plants specific comments are made as to certain factors to look for, but in general it may be said that a relatively large quantity of dyestuffs is needed for dyeing a pound of wool, as the pigment in a given piece of root or bark may be very small. As a basic rule it may be suggested that the following quantities are minimal for a pound of wool:

Bark — one peck finely chopped

Leaves — three-fourths peck, dried

Hulls — one peck

Flowers — 1½ quarts of dried or 1½ pecks fresh flower heads

Roots — highly variable

Now follows a partial list of the colors that may be produced from the plants, with the exact finished color depending on the type of mordant used, dilution of the extracted pigment, the number of dippings, and other factors.

Violets	*Reds*
Dandelion	Barberry
Wild cherry	Cranberry
Maple	Pokeberry
Grapes	Seaside rose

Blues
Elderberry
Bayberry
Grapes
Horsebrier berries

Yellows
Apple
Barberry
Mullein
Hickory
Bracken
Sumac
Goldthread
Celandine

Beige and Tan
Sassafras
Lichens
Poke leaves
Walnut hulls
Sumac
Oak bark

Browns and Blacks
Cherry bark
Birch
Black walnut
Blue flag
Juniper
Pokeberries

Preparation

After collecting whatever plant is to be used, the material should be cut into small pieces and pounded, to make it possible for the pigment to separate itself. Many materials so prepared may be dried and stored for winter use.

The macerated pieces of dye material may be put into a pot on the stove (an enameled pot is best) and covered with water, brought to just below the boiling point, and then simmered for some hours until all the color has been extracted. The length of time would depend on the kind of material, bark taking much longer than berries. All attempt at extraction should be discontinued as soon as it seems as if no further color is appearing. In fact, some dye plants, if cut or broken into small pieces, and allowed to stand in water over night will yield a strongly colored solution. To eliminate the necessity of straining fibers and roots from the liquid, the plant parts may be boiled in a loose cheesecloth bag which is easily removed when the color is extracted.

In this connection it may be noted that after extraction certain colors may be combined, for the production of various soft shades and colors, but the knowledge of what to combine would come only through long experimentation. This is especially true because the color of the extracted dye very often has but little relation to the final color of the finished product.

Dyeing

In discussing the actual dye process comments here concern wool and wool alone, for other materials do not seem to have an affinity for natural dyes in the same degree as wool. Silk, especially, is unpredictable, while cotton requires additional processes for mordanting. Again, it is wool with its off-white natural color, that absorbs dyes in a way that makes it peculiarly beautiful when dyed. Hence we shall in these comments leave the formulae for cotton, linen, and silk for future exploration and suggest that the beginner stick to the dyeing of wool.

Dyeing would be an utterly simple process if one could count on all plant colors being "fast" and washable, but almost none are, and so something must be done to the wool, either before or after dyeing, to fix the color. This is the *mordanting* process which can be done either before, during, or after actual dyeing, the time of doing, having considerable to do with the intensity and fastness of color. Here is another place for individual experimentation, and here again we shall, for simplification, discuss one method, the pre-dyeing *mordanting*.

OAK GALLS

The materials used in mordanting are alum, chrome (potassium dichromate), copperas (ferrous sulphate), and tannic acid (commercial or from a natural source of tannin such as oak galls or sumac leaves). By using different mordants a variety of shades and sometimes even different colors may be obtained from the dye.

Alum Mordant

This is the most commonly used of all mordants and, used in combination with cream of tartar, brightens the wool. For each pound of wool to be dyed (*all* recipes given here are based on 1 lb. of wool) use:

3 or 4 ounces of potash alum to 1 ounce of cream of tartar. Dissolve these in about 4 gallons of cold soft water. Put on stove and as the water warms up immerse the wool (after it has first been wet in water and squeezed dry). Now heat gradually to boiling, stirring and turning the wool all the time. Boil gently for one hour, adding more water to keep the proportion of liquid to wool always the same. Remove from fire and let wool stand in the water overnight. Next morning remove wool, squeeze dry, roll in a towel and put in a cool place ready for actual dyeing.

Chrome Mordant

Dissolve in 4½ gallons of soft water, ½ oz. potassium dichromate and proceed as for alum mordant above.

Dyeing

And now we come to the actual dyeing itself.

1. Dye the full amount of any possible yarn or cloth of each color needed for a piece of handicraft at one time, instead of trying to match the color by a second dyeing. Vegetable dye materials vary so much that it is almost impossible to duplicate colors exactly. Anyone who has tried to match a given yarn or cloth will realize that this same statement is true of commercial dyes also.

2. If work is done in the summer time, there will be less damage from spills if the work is done outdoors.

3. To obtain colors lighter or darker, vary the proportion of dye to water.

4. Wool should never be boiled hard or stirred around very much, while being dyed, as it may cause shrinkage and matting.

For the operation itself, the formula for the color and material can be used as listed under each plant. The dye-stuff which has been prepared as above is put on the stove cold. The material to be dyed is first rinsed well (to remove excess mordanting solution) and squeezed dry. If skeins of wool, they may be very loosely tied and dipped and removed on sticks or wooden forks. Keep dipping and dyeing or stirring gently in the now-boiling dye, until the material has reached the desired color. Then remove from the heat and allow the material to cool in the dye water. Then lift it out, carefully squeezing out excess mixture and rinse in clean water until no color comes out with the rinse. Now dry in suitable place, and the job is done.

Sometimes for certain colors (and this is for experimentation) a hot soap bath after dyeing will brighten and further set the colors.

Dyeing Procedures

Beyond this somewhat standard method in dyeing there are other possible orders of procedure.

1. The wool may be boiled first with the dye, then the mordant can be added to the same dye bath.

2. The wool may be boiled with dye and the mordant in the same bath together.

3. The wool may be first mordanted, dyed, and then mordanted again to insure the maximum of fastness.

Each of these procedures will produce different color effects, and here is where experience based on experiment, comes into the picture.

In summation it may be said that much of the success of the whole process depends on the care used in mordanting, while the character of the color depends upon the nature of the mordant. The chrome mordant mentioned tends to produce a golden hue with some materials, tin brightens the colors and iron darkens the color. If one were interested in using entirely botanical material for both dyes and mordants, there are possibilities of using oak galls, wood ashes, sumac, hemlock, sorrel, and similar plants to provide the necessary chemical changes that will enable the dye to bite (Latin-mordere) into the wool, all of which and many other possibilities, are suggested in some of the more detailed books on dyeing.

And now to a discussion of some of the native dye plants. It would be more complete and interesting if the uses and methods connected with such well-known dyes as logwood, indigo, madder, and saffron could be explored, but as this survey is purely concerned with 100 plants of the Northeastern United States we will confine comment to that group.

PLANTS PRODUCING DYE IN SHADES OF RED

Pokeberry

Pokeberries to produce the best red color should be fully ripe. They may be soaked overnight in water or simmered slowly until all juice is out and then filtered to remove pulp. If berries are wanted for later use they may be carefully dried for winter use, being on guard to avoid mold. The dyeing process should follow regular procedure as outlined, except that a special mordant can be prepared by substituting a half gallon of vinegar in the place of alum. This will produce a color similar to that made with cochineal.

SHADES OF YELLOW

Apple bark

Apple bark produces shades varying from dull yellow to pure gold or brass. The bark may be used either fresh or dried and different varieties of apples will produce different shades of yellow. A peck of bark (to a pound of wool) would be required. Soak the bark over night in a couple of gallons of soft water. Boil for several hours, replacing evaporated water to keep a constant quantity. Strain and add soft water to make 4 gallons. Cool until lukewarm and then immerse material and boil 30 minutes.

Celandine

Boil up the roots using recommended mordant. If just the right quantity of plant material is used a true yellow will result.

Mullein

Dyeing with the juice extracted from mullein plants and with alum-mordanted wool, is said to afford a "lovely bright greenish yellow that calls to mind the sunshine."

Goldenthread

With this plant not only are the roots used for medicine but also for their production of a fine yellow color as a dye. Alum mordant.

Sumach

The use of a half peck of sumac berries to produce a dye solution, and mordanting with alum will produce a yellow color. See note about its use for producing black.

Common Barberry

The wood and inner bark of the tall wild barberry bushes are a bright yellow, and make a good, strong, yellow dye by cutting the bark finely, covering with soft water, and simmering until the color is extracted. The resulting dye may be used without mordanting as it is a "fast" or substantive color.

Hickory bark

A dye prepared from the macerated inner bark of hickory produces on wool a good yellow color. Again other mordants and amounts of bark will produce variations.

SHADES OF BLUE

Elderberry

Both the berries and leaves have been used as sources of dye, the berries after processing produce in wool a lilac-blue dye while the leaves give a green color.

Bayberries

When wax for candle-making is being extracted from bayberries by boiling, the remaining water will dye wool blue, without mordanting. Leaves of bayberry dried and then boiled, will make (according to quantity) a fast yellow or yellow brown. Mordant with alum.

Horsebrier

It is a little hard to think of this plant as having much use, after one has tried to walk through a thicket where the vines are growing and has been torn by the briers, but the blue berries are reputed to be an excellent source for dyes in various shades of blue, navy, purple and violet. Mrs. Kierstead has this to say in her book. "For shades of blue, dye the goods for a short time in the dye, then to change the color to blue, lift the goods out and add table salt in considerable amounts; the more salt used, the darker the blue color will be, and finally a deep midnight blue may be obtained. The presence of salt, by increasing the temperature of the boiling bath, also helps make the goods fast to washing."

PURPLES

Chokecherry

The use of the roots of the wild cherry tree when used with alum mordant will give a reddish-purple color, two pounds of roots being required to a pound of wool. The bark of the tree, on the other hand, produces a range of colors from yellow through orange and tan.

Dandelion

It is recorded that the Scotch have used the roots of dandelion to obtain a magenta dye for weaving into their tartans, but in this country without mordant the roots, processed fresh, give shades of yellow of good fastness.

TANS AND BROWNS

Sassafras

All authorities agree that the pinkish-brown result of dyeing with the root-bark of sassafras is a beautiful and desirable color. Twelve ounces of the dried root-bark, which has been soaked overnight, is boiled for 30 minutes, strained and water added to make the usual 4 gallons. After boiling the material in the dye for 30 minutes and without rinsing, it is put in a boiling bath of water to which 1/6th ounce each of potassium dichromate and acetic acid have been added. Boil 10 minutes, rinse and dry the wool.

Oak bark

The dried bark may be boiled and used as a dye with the formula as given and (depending on variety of oak and concentrate) produces colors ranging from yellow tan to light brown.

Sumac

All parts of the sumac may be, and have been, used as dye material. The berries (about a peck) will produce a light tan color. Addition of roots or branches produces shades of grays up to black.

Lichens

A recent and authoritative survey of the genus LICHEN and its uses among Arctic peoples indicates that as many as 40 or 50 species are possible sources for dye. Here is a field for interesting experimentation with the more common forms of rock and tree lichens that one sees so often. The lichen is a "substantive" dye and requires no mordanting for use with wool. Experiments show that forms growing on stones generally give better colors than that from trees. Two methods are given for dyeing, the colors resulting varying from yellow to brown.

1. Put the lichen into a large pot and fill to the top with cold water. About one pound of lichen to one pound of wool is required. Bring this up to the boil very slowly and let it simmer for two or three hours and then get cold. Next day, put the wool, which should be thoroughly wet first, into the pot, and boil it all up together until the desired depth of color is obtained. The wool should not be taken out until it is cold. Then wash. The loose lichen shakes off easily.

2. A layer of lichen at the bottom of the pot, then a layer of wool, then a layer of lichen, and so on until the pot is nearly full. Fill up with cold water and put on the fire to simmer for some hours until the required depth of color is reached. It can be taken off at night and put on again next day. The color will be very fast. Wash well.

Brown colors are obtained from forms of the genus Parmelia, yellow colors are obtained from various species of Ramalina.

Birch

Two pounds of fresh birch leaves used as dye in accordance with the routine procedure will yield a yellow dye while the dye from bark gives a reddish brown.

Blueberry

As most housewives know, who spill it, boiled blueberry juice is "very permanent." Its color used as a dye is a dark blue but if processed with oak galls will give a dark brown. One pound of berries to a pound of wool.

Blackberry

As with so many plants, the colors obtained from the fruit and from other parts of the plant, vary widely. The young green shoots will, if mordanted with iron, produce an almost black color, while the berries mordanted with alum will give a pleasing bluish-grey.

Black walnut

As may be suspected, the color obtained from walnut hulls is a pleasing shade of brown. Hulls may be collected and stored in a jar with water for a considerable period to extract the color. Actually no mordant is needed, as the color given to wool in this mixture with boiling is quite fast. If wool previously dyed blue, is boiled in walnut juice a fine black color may be had, or it is possible to produce black alone from walnut trees by gathering the autumn leaves as they fall from the tree and do the dyeing in this manner: place in a kettle a layer of walnut leaves, then one of wool, one of leaves, and so on. Cover with water and boil 12 hours. Cool over night, pour off juice, prepare fresh series of layers of leaves, pour on old juice and repeat boiling. This dyed yarn, after hanging for a few days and washing, will be a pleasing black.

Juniper (Red cedar)

The bark, berries and twigs are all suitable for dyeing wool. Berries are commonly used and produce with this special recipe, a good khaki color. For a pound of wool proceed as follows:

2 quarts ripe juniper berries
2 ounces potash alum
3/4 ounce ammonium chloride
1 ounce cream of tartar
1 ounce copper sulphate
1 ounce copper acetate

Dissolve the alum, ammonium chloride, cream of tartar, and copper sulphate in 4 to 4½ gallons of soft water. Put in the wool, wet thoroughly and squeeze out the water, boil for 1 hour, and let stand in this mordanting liquor until cold, then rinse. Break up the berries, tie in a cheesecloth bag, soak in water overnight, then boil for 1 hour and add enough cold water for the dye bath. Immerse the mordanted wool in this dye extract, boil for 1 to 2 hours, and remove. Add the copper acetate to the dye bath and when dissolved, return the yarn or cloth, and boil for 15 to 30 minutes longer. Rinse and dry.

"The Lord has created medicines out of the earth;
And he that is wise will not abhor them.
And of his works there is no end;

ECCLESIASTICUS 38-4

Chapter VI

REMEDIES ON THE ROADSIDE

REMEDIES ON THE ROADSIDE? Yes, there are plenty of them if we would but think about them and hunt them out. A visit to the druggist and a look at his shelf of "materia medica" would produce a list of names not unlike a catalogue of roadside plants. The problem is largely one of knowing when and how to use this bounty of nature.

It is impossible in a book of cursory comments like this to explore all the diseases of man or to indicate the cures. Nor is it wise for the discomfited traveler to chew the leaves of the overhanging trees to search for a cure. The family doctor surely has his place in modern life just as does the grocer or the clothier. Hence this chapter does not purpose to suggest wide experimentation with medicinal plants but is written to increase the knowledge of the reader to the point of a greater appreciation of the uses of plant material.

In the early days of this century when there was a general disgust with, and neglect of all cures which were "nonscientific" and a growing belief that medicine and chemistry were closely allied, the efforts of the old time herbalist were completely disparaged. But fifty years and the discovery of penicillin have brought about a new appraisal of empirical medicines. Scientific research, taking into account the medicine of ancient China and other early cultures, has tempered the prejudice against plant medicine, while many "folk cures" have been found to be scientifically sound. Since the early findings of the value of penicilin as an antibiotic, there has been a steady search for similarly medically valuable qualities which might be found in higher plants. Thus it has been discovered that many ancient therapeutic plants not only have had curative but preventative properties as well.

In the Agricultural Yearbook, 1950-51, there appeared articles by Schaffer and others, which told of the bacteria-killing properties of some of the plants mentioned in this book. That we are on the road to a use of plant medicines on a scientific basis seems to the writer to be an unquestioned fact.

Here in our own Northeastern America the first inhabitants knew most of the values of plants long before we arrived says Janet G. B. MacCurdy in the HERBARIST (1941).

"Among the members of the Gay Head and Mashpee groups in Massachusetts it is believed that a female deity, "Granny Squannit", controls the plants and that in earlier times medicine men and women offered food to "Granny", in return for which she guided them to places where they might find certain rare plants.

This knowledge was considered as personal property and persons possessed of power to heal the sick either through magical practices with the use of herbs, or those affecting cures with the use of herbs, were held in high esteem by members of the tribe."

The author goes on to say that wherever white men came in, the valuable plants disappeared and that the succeeding generations had less and less knowledge and had to have recourse to questioning "Granny Squannit."

"In gathering plants it was important not to take more than the person expected to use. In most Algonkian tribes it was customary for the gatherer to offer a small amount of tobacco to the spirit of the plant. This was done by digging a small hole toward the east side of the plant and placing tobacco in the hole. The first plant of a species sought by the "doctor" is not picked but he leaves the offering and then walks on. Presently . . . he sees a number of plants of the desired species and takes what he needs.

Another recent study of the plant medicines of the Indians indicated that for a Redman with a stomach-ache there were at least 100 plants from which he could choose. Outstanding among these was the root of the sweet flag, which the Indian considered a cure-all. No preparations of sweet flag were required, as the Indian simply chewed the root for a long time. Teas brewed from the dried leaves of berries of wintergreen and other plants were used for various illnessess, while our own modern medicine uses the same plant just as wisely under the name "Oleum Gaultheria." The Indian knew also that the oil of the Sassafras tree was good for fever, as were extracts of one variety of goldenrod, and a preparation from the plant which we call boneset.

The Indian even had discovered that the dried powder of the puff ball was useful for stopping the flow of blood, the spores being dusted on the bleeding wound. They knew, too, about similar properties of the bark of the staghorn sumac, which, because of the tannin it contains, has a puckering effect on the skin. And so one could go on to trace many of these medicines to their Indian origin, and he would be surprised to find how many of them are still useful.

"Dr. Thomas Lewis of the Revision Committee of the U.S. Pharmacopoeia, is a strong advocate of research into primitive cures. He believes that: 'Even in this age of synthesized drugs, invaluable therapies may be lost to medicine if we don't make an exhaustive exploration of those cures which primitives apparently have always been able to find in plants.' "—Don Romero, "Black Magic for Men in White," in *Nature Magazine* (June, 1953), pp. 301 *et seq.*

Now today we of the world of chemistry can put to simple use some of this knowledge, permitting simple successes in herbal medicine to lead us on to a greater study, appreciation, and use of plants. Countryfolk generally know, for instance, that the eating of blackberries tends to stop dysentery, and that sphagnum moss is fine for an emergency bandage on a wound. Many know that an extract of the bark of the wild cherry tree is fine for cough syrup, and with an observing mind we can learn many simple facts that will be helpful, as we come to appreciate nature's remedies.

It should be understood in this discussion that the author is not making any effort to build up a case for the "herbalist." Modern herbalism, it is true, is quite different from the herbalism of medieval times, but like all other "isms" it may be that at times it builds up a great case around a few simple facts. The all inclusive statement of the modern herbalist that "medicines made exclusively from plants are the natural cure for human disease and ill health" is a sweeping one, and one cannot but feel that the interest in these cures is quite akin to the unquestioning acceptance of the claims of the school of organic gardening. Nor should we discount in this consideration the value of faith as an assistant in herbal cures, nor the gradual incorporation into herbal medicinal theories of a knowledge that has been thousands of years in the building.

The best view today ought to agree with the comment of Dryden:

Better to hunt in fields, for health unbought
Than for the doctor for a nauseous draught.

The knowledge of plant medicines has been long in building. Take, for instance the case of the very old "doctrine of signatures," in which the shape of a plant, as related to the human body, indicated its possible curative value; for several centuries such a theory held sway in the field of medicine; until by trial and error and scientific analysis, the theory was completely disproved. A remnant of this and of many folk tales of medicines still persists, however, and the plants indicated have the value only of a faith cure.

The author well remembers that when he was growing flowers commercially that there was more or less constant demand for the shipment of the leaves of fragrant violets to customers who wanted them for medicinal purposes, presumably for the cure of cancer. There is no evidence whatsoever to indicate any such value, but the belief in such a cure has remained in spite of modern medicine, and it is understandable that the desperate person goes on the theory that one can try anything, once.

The true "herbalist" doctors, of which there are a few in Western civilization (including especially England) and many more in the East, tend to be serious practitioners and resent any implication of quackery. They are as apt (as is the regular doctor) to give medicines to the patient without naming them, and undoubtedly keep a great deal of their knowledge as a mystery. One such doctor writes as follows: "An herbalist who really understands herbs treats the individual patient, and the art of herbalism lies not only in choosing the right herb for each of these diseases but also in selecting the correct one for each individual sufferer from a particular disease.

In sum, then, we can note the words of an orthodox practitioner, Dr. Douglas Guthrie, Lecturer in the History of Medicine of the University of Scotland who said, in a lecture in Boston in 1954, that *Folk Medicine is not to be despised.* It should be helpful, therefore, for all of us to know the accepted virtues of our roadside plants.

Following the general practices of most books on herbal medicine, the author would call attention at this point, to the fact that not only does a great deal have to be known about the specific properties of an individual herb before using it, but also about the specific disease as well. In offering on the following pages a list of the medicinal plants of our area from among the one hundred plants of this book, note: that not only must all herbal medicine be administered with unquestioned knowledge of both disease and plant but that care also be taken to properly extract and preserve the medicinal properties in the plant.

A few words then on the method of providing a supply of herbs for medicinal plants. It is obvious that to have material for use throughout the year there must be a careful preparation and collection at the right *season* of the year. In the case of many plants the vital properties are usually most highly concentrated in the season just before the plant comes into bloom. When gathering the particular plant parts which are to be dried or preserved, as indicated in the list below, it should be remembered that mold is apt to develop if the plants are picked with moisture on them. Gathering should be delayed until the sun has dried all dew and after collection, the plant parts may be tied together in bunches and hung in a warm and well ventilated attic. An air current blowing around them will help to promote drying. After complete drying, the leaves or plant can be ground up, or pounded into an almost powdery condition and stored in absolutely dry jars which have previously been washed clean with soap and water, being certain to stopper tightly and to label each jar carefully.

In the more detailed comments which now follow, a number of plants will be discussed, which have even a minor or possible, medicinal value. In not all cases would it be ordinarily convenient to extract the active principles from these plants, but perhaps a consideration of the qualities of these wildings will make us more appreciative of the many values of plant life.

THE APPLE

One of the first maxims that this author learned as a boy, was the familiar one, *"An apple a day keeps the doctor away."* There is another form to this which says, *"To eat an apple going to bed, will make the doctor beg his bread."* Generally speaking, we are well aware of certain values in the apple, as shown in our use of apple-sauce with roast pork and rich dishes. One authority says that a good ripe raw apple is one of the easiest substances for the stomach to deal with, and he also points out that two or three apples taken at night, either baked or raw, are admirably efficient to alleviate constipation. Certainly one might say that considering the aromatic quality of the apple and its commonly known values, this is first among our possible and pleasant medicinal plants.

SASSAFRAS

Among the many common names of this plant, one of the first is that of the *ague tree,* which indicates its early use for medicinal purposes. In fact, it was the search for the sassafras as medicine, which caused explorers such as Gosnold, to come first to the shores of America. The aromatic qualities of the sassafras are still much used in medicinal preparations, although the value of the oil for curative purposes is doubtful. The part of the tree now used is the bark of the root, both epidermis and cambium. The roots are distilled, and the product sold as an oil which goes to flavor drinks and medicines.

DOGWOOD

Early explorers in the United States used a decoction prepared of the dry bark of dogwood in place of quinine. Also, in those early days, the peeled twigs of the dogwood were used as toothbrushes, because it was thought the chemical content would whiten the teeth.

SWEET FLAG

Of all the one hundred wild plants discussed in this book, perhaps the plant of oldest medicinal use is the sweet flag or *Calamus*. Accounts of its use go back to Roman times or even earlier, and there is good evidence of its value in many ways. These medicinal values were long known to the American Indians, and it is said that they used it without hesitation for a great many ills. They considered it a very powerful medicine, and a dose was a piece of a root of the length of a finger joint. The Indians apparently used the root in its unprepared state, and chewed it for every kind of difficulty from toothache to kidney trouble. They used the juice to heal burns, the dried root to be snuffed up the nose for catarrhal conditions, and the pulp of the root was applied as a poultice.

Among our early settlers, the root stock was dug up in the spring, washed, and when dried was chewed as a stomachic, the unpeeled root being more efficacious than the peeled. Under present-day conditions it is still recommended for an upset or sour stomach, and the following method of preparation is given:

One teaspoonful of dried root to a cup of boiling water. Drink cold one or two cupfuls a day. For administering to children, the root may be grated and the pulp put in a glass of water with a little sugar.

The roots of this plant may be gathered in the autumn or spring. The rhizomes or fleshy roots should be cleaned of the fibrous parts, and dried with moderate heat.

One of the interesting early uses of the sweet flag was the utilization of the pleasant smell that it gives out under heat. For this reason the floors of cathedrals were often strewn with sweet flag, and our New England forefathers used it for the same purpose in the home. There are still minor uses for the extracted oil in industry. Altogether, this is a plant worth knowing.

SHIELD FERN

Among the wild medicinal plants listed by the Department of Agriculture are several forms of the wood ferns, especially the Male or Shield Fern, (Aspidium Filix-mas) which has for centuries been known for its value as a remedy for intestinal worms. The part used is the root stock, which is collected in late summer. One would probably go to a doctor for the cure of such a condition, but it is interesting to know that the plant has these medicinal qualities, as well as a value as food when the fronds are young. One of the most interesting facts about ferns is that for many years the seeds (spores) of this and other ferns were thought to confer invisibility on the eater. Hence we find in Shakespeare, in Henry IV, this statement, "We have the receipt of fern seed; we walk invisible."

SHEPHERD'S PURSE

One of the prime pests of the farmer is this weed which has been introduced into our country and which has followed man in his migration. All herbalists agree that it is a fine remedy for bruises, and there is considerable evidence to show that in times past, it was found to be exceptionally good when taken internally, to stop bleeding. Recipe books usually indicate that it is best taken as an infusion, which can be made by boiling an ounce of the leaves of the plant in twelve ounces of water, reducing by boiling, to half a pint, then taking a wine glassful as a dose.

CHICORY

In addition to the roots of this plant being used in times past as a substitute for coffee and as a present day additive to coffee in the South, the chicory has long been used medicinally. Probably for the same reason that dandelion (its cousin) has always been considered a good vegetable to eat in the spring for its tonic powers, so the chicory contains the same basic ingredients. Several authorities are agreed on the value of distillation of the chicory flowers to allay inflammation of the eyes, but from the standpoint of this writer the main values of chicory are as an ingredient in coffee and as a spring green when blanched.

GOLD THREAD

The various common names of Coptis indicate its use in medicine, for Canker Root and Mouth Root are both given as frequently used names. As an ingredient of medicine it is still commonly collected and sold to druggists. Herbalists consider it as a pure and bitter tonic with highly astringent qualities which might be useful in curing canker sores. One book suggests that a decoction of the root of Gold Thread will cure the taste for alcoholic beverages. To prepare, steep a teaspoonful of the ground-up, dried, roots in a cup of boiling water, for about half an hour and take a tablespoonful of this mixture several times a day.

WINTERGREEN

The use of wintergreen for the flavoring of medicine is an old practice, and in some parts of the country the wintergreen plants are collected for distilling oil of wintergreen.

BLACKBERRY

If you are one who likes to make the fullest use of native plants it will be useful fun to prepare a supply of the medicine which the blackberry offers, for the entire plant contains an active principle which is about the best cure known for diarrhea. The roots of blackberry (and leaves, as well) are dug and dried and stored until wanted. To prepare for use, take a teaspoonful of crushed root and stir in a cup of boiling water. When cooled, drink one or two cups a day until condition is remedied. (To which we might add that the eating of a large quantity of the fresh fruit will have about the same effect).

SKUNK CABBAGE

The very name of this plant and its offensive odor in the spring, is enough to repel anyone from any possible use of it, but the pharmacist uses an extract of the plant under the name of *Dracontium*. It is a narcotic and stimulant and should not be used except by a doctor's prescription.

NARROW-LEAVED DOCK

The use of the root of this plant as a spring tonic and stomachic goes back many centuries and the plant is still listed among the recognized pharmaceutical plants of the U.S.A., and for those who believe that the old herbalists "knew a thing or two" it might be interesting to dig and dry a dozen roots of dock and make use of them as occasion requires. Some stimulation of the stomach may well ensue and undoubtedly vitamins will be added to the diet, to say nothing of a patch of ground cleared of weeds. Instruction books suggest that a teaspoonful of the dried, cut-up root be steeped for a half-hour in a cupful of plain boiling water, the mixture to be drunk at the rate of 2 cupfuls per diem.

DANDELION

There is a great body of literature on the dandelion, which is too long to quote here. For many centuries the use of the dandelion has been common in medical practice. The Arabian physicians of Spain who did much to promote modern medicine, mention the plant as Taraxacon, which is a Greek word meaning "remedy for disorders." The part of the plant used for medicine was the juice, which in winter, after frost, is quite sweet. To a great extent the use of the dandelion plant in medicine is now, however, discontinued. There are probably many benefits which come from the juice, and people from the Mediterranean countries value the dandelion highly to use in the spring as a green with tonic properties.

SPHAGNUM MOSS

Sphagnum Moss is commonly found growing in swamps and is easily identified. It is probable that the Indians in this country used Sphagnum Moss for many centuries, and it is only recently that medical men have realized some of its advantages. The dry moss has the quality of absorbing at least twenty times its own weight in water and for that reason it is extremely desirable material to use for cases of bleeding. It is a good thing to know this when one is traveling in the wild, and needs to bandage a wound, because not only does the Sphagnum Moss act as an absorbent, but it seems to have antiseptic qualities, probably accounted for by the presence of iodine. Other chemical conditions found in a sphagnum bog have healing qualities and one should not hesitate to use this moss when it is found growing naturally in a bog. There are many other uses for Sphagnum which fall under other headings, but its emergency medicinal value should not be forgotten.

JEWEL WEED

As a drowning man will "clutch at a straw" so the sufferer from poison ivy will try almost anything. The preventatives and remedies for poison ivy are many and seem to affect persons differently, so the author would not want to recommend any one as being perfect. There are, however, quite a few references to the use of the juice of the jewel-weed as a check to infection with ivy. Certainly no harm would be done in using it and if one were confronted with the knowledge that contact had been made with poison ivy, it would be the part of wisdom to apply jewel-weed juice directly from the fleshy stems of this common weed which, fortunately, is often found growing not too far from ivy.

SUMAC

One authority gives a recipe for a gargle to be made of the berries of the red-fruited sumac — the same ones that make a delightful summer drink. A bunch of berries are boiled in a pint of water for a half hour, then are added a few leaves of dried sage, and a little ground cinnamon. After straining this can be used as a gargle as needed.

MULLEIN

With the dried foliage of mullein and/or of BONESET HERB a cough syrup is suggested by the experts in herbal medicine. A handful of leaves are simmered for about 20 minutes, stirred and strained. To a cup of liquid add a little over a cup of brown sugar. Stir until sugar is dissolved and bottle. This may be sipped ad lib to alleviate the distress of a cough. Or the same mixture may be boiled until it thickens and poured onto a pan and cut into lozenges while cooling as coughdrops.

IRISH MOSS

The Irish Moss which one finds along the shores of the Atlantic ocean is said to be the basis for a fine drink for lung troubles. This is probably due to the mucilaginous nature of all of the seaweeds. One recipe for preparing the Irish Moss for medicinal use is as follows:

Wash clean a few pieces of Irish Moss; put it in a cooking pan and pour over it two cups of boiling water. Set where it will keep at the boiling point, but not boil, for two hours. Strain, and squeeze into it the juice of one lemon. Sweeten to taste. If the patient cannot take lemon, flavor it with wine, vanilla, or nutmeg. To be used as desired for a "lung condition."

IRISH MOSS

Again, the gelatinous qualities of Irish Moss make it a natural ingredient for a home-made Hand Lotion. B. C. Harris of Worcester suggests mixing a teaspoonful each of Irish Moss and Tragacanth flakes in a pint of boiling water and adding when cool a teaspoon of borax and lastly an ounce each of glycerin and rubbing alcohol.

HONEY

The probabilities of finding wild honey along the roadsides is not too great but at least honey can be easily come by and it is interesting to know of its many values. A doctor (W. Schweisheimer) has this to say about honey in a recent issue of the RURAL NEW YORKER: "The healing properties of honey were never forgotten by folk medicine. Recent reports stated that honey was a favorable influence in cases of suppuration of the bladder and for kidney diseases, since it stimulates the activities of kidneys as well as of the bowels. Honey has a destroying effect on bacteria which probably comes from its content of dextrose. For this reason it is not inclined to form mould.

"Honey has been recommended by many doctors for its generally loosening, cleansing and strengthening effect. One or two teaspoonfuls taken before bedtime may relieve insomnia.

"Honey is quickly and almost completely digestible. Therefore it is a quick-acting source of muscular energy."

The fact is, the suggested ability of honey to kill germs is well known and it is said that if a germ-laden spoon were plunged into honey overnight, there would be no trace of the germs in the morning. The use of honey in old recipes for the dressing of wounds would indicate a long time knowledge of this factor.

BEARBERRY

An extraction of the leaves of this plant are reputed to have considerable value as a diuretic and tonic. The dry green leaves are picked in the fall, thoroughly dried, and reduced to powder. For use they are soaked in a little brandy or grain alcohol (½ tsp. as a dose) and this solution is added to a cup of hot water and taken twice daily as a "mild disinfectant to the urinary tract."

WILD GINGER

As is true of the ginger of commerce, which belongs to an entirely different species of plant, the wild ginger is valuable as an aromatic bitter and is used medicinally for flatulency. The small roots of the plant are to be dried and when needed, prepared by adding a teaspoonful of the granulated root to a pint of boiling water. Of this mixture 2 tbsp. may be taken at a time as long as required.

WITCHHAZEL

The value of the witchhazel as an astringent is extremely well known and the plant is generally so common in the woods that its use from the wild is not impracticable (at least on an experimental basis). Gather freshly-cut twigs and leaves and soak them in water for a day, using about two parts of water to one (in bulk) of cut-up plant parts. This solution may be reduced by distillation about 25% and to this extraction, alcohol added at the rate of 1 pint of alcohol to 5 parts of the distillate. This is to be used as a stimulating rub for sore muscles.

WHITE PINE

The medicinal uses of the bark of the white pine have long been recognized. Medicinally it is an "expectorant" used mostly in cough remedies. The inner bark and sprigs of new growth are the parts used. The brew is usually mixed with extracts of wild cherry, sassafras, and honey to secure the best results.

WILD CHERRY

The commonest ingredient in many cough syrups is wild cherry. The bark is collected in autumn, the outer layer removed and the inner bark then carefully dried and preserved. Do not use bark from very young or very old branches and do not keep bark too long as it deteriorates with age. A teaspoon of this bark infused with boiling water may be used separately as a drink but preferably added to other ingredients. The fermented juice of the very ripe fruit bottled after fermentation ceases, has been widely used as a cure for diarrhea.

BONE-SET

It is doubtful if the use of this plant would set any broken bones but references to many herbals indicate that it was used for many years as a pain-reliever for conditions of the limbs or muscles. The HERBALIST advises its use at the rate of one teaspoon of dried foliage to a cup of boiling water. Strain and take at the rate of a teaspoonful five times daily.

RED CLOVER

It perhaps is no coincidence that the four-leaved clover is a sign of good luck, for the familiar large-flowered red clover of the farmer's field is valuable in more ways than one. The food value of the plant for cattle and its incidental nitrogen-producing qualities in the soil are well known, while most children (and the bees) know of the hidden sugar in the florets.

Additionally there seems to be considerable medicinal value to an infusion of the flowers, especially when used fresh. It has a demulcent action which loosens coughs, lessens difficult breathing, and is especially recommended in cases of whooping-cough. Boil up and strain an ounce of flowers to a pint of syrup sweetened with sugar or honey. A teaspoonful twice a day provides a safe and reliable medicine.

CHARCOAL

One doesn't have to chew somebody-or-other's Charcoal Gum to get the values that the use of just plain charcoal can effect. Of course this may be a bit messy and so a tablet or gum from the druggist may be easier. But if one is troubled by flatulence and is out on a camping trip, just remember that a gaseous condition in the stomach and/or a bad breath may often be relieved by taking a small quantity of powdered charcoal.

In the home it is reported that charcoal placed near "high" game will take away the smell, as it will of other unpleasant kitchen odors. There are many other uses for charcoal in such fields as gardening, but they all derive from the natural tendency of this product to absorb gases of all kinds, thus "sweetening" almost anything.

Not every one of our hundred plants has acquired a reputation for its curative properties nor can we here explore in detail the less important of the plants. Below in alphabetic order is listed the reputed medicinal virtues of such of our plants that have not been more fully discussed in the preceding pages, but which have long been used in herbal medicine.

Arctium minus — BURDOCK — Skin diseases — Psoriasis

Asparagus officinalis — ASPARAGUS — Diuretic

Atriplex — ORACH — Used in treatment of hysteria

Berberis vulgaris — BARBERRY — Cooling and astringent

Asarum canadense — WILD GINGER — Aromatic, used for flatulence

Betula lenta — BIRCH — Used in Birch Beer — Oil of wintergreen from bark

Ceanothus americanus — NEW JERSEY TEA — Substitute for tea — Gargle

Chelidonium majus — GREATER CELANDINE — Caustic for application to warts

Comptonia peregrina — SWEET FERN — Astringent and mouth wash — Worm cure

Crataegus — HAWTHORN — Kidney ailments and sore throat

Equisetum fluviatile — HORSETAIL — Mild diuretic

Fragaria virginiana — STRAWBERRY — Leaves valuable in treating dysentery

Galium verum — OUR LADY'S BEDSTRAW — Healing — used in lotions

Gnaphalium obtusifolium — EVERLASTING — Healing in local applications

Iris versicolor — BLUE FLAG — Liver and bilious condition

Juglands species — WALNUTS — Itching and eczema

Juniperus communis — JUNIPER BERRIES — Kidney ailments

Lycoperdon gigantea — PUFFBALL — External coagulant

Malva — MALLOWS — Demulcent and emollient

Mentha species — MINTS — For flatulency and flavoring

Nymphea odorata — POND LILY — Scalp ointment

Phytolacca americana — POKEWEED — Cathartic

Populus species — POPLAR — Buds used in cough syrup

Rosa rugosa — JAPANESE ROSE — Seed pods (Hips) are source of vitamin

Rumex acetosella — SHEEP-SORREL — Healthful when used green

Sambucus canadensis — ELDERBERRY — Diaphoretic

Sanguinaria canadensis — BLOODROOT — Expectorant and stimulant

Solanum dulcamara — NIGHTSHADE — Relief of skin irritation

Solidago species — GOLDENROD — Aromatic — mildly stimulating

Tilia americana — BASSWOOD — Promotes perspiration — relieves cough

Viburnum trilobum — HIGH BUSH CRANBERRY — Currently used as Anti-spasmodic

DANDELION **NARROW-LEAVED DOCK**

Though nothing can bring back the hour
Of splendor in the grass, of glory in the flower:
We will grieve not, rather find
Strength in what remains behind.

William Wordsworth

Chapter VII

WILD PLANTS IN AND OUT OF THE HOUSE
THE WILDINGS AS DECORATION

PART I — *Plants Indoors*

JUST AS THE STYLES OF WOMEN'S DRESS seem to go in circles which repeat themselves at intervals, so, too in styles of home decoration. In Victorian days, the use of dried wild flowers of all kinds was extremely common, to be followed by a period when they were highly despised. Now again, it seems to be fashionable to use bouquets of dried grasses, weed seeds and other material, (often and regrettably dyed or sprayed). So, for the city dweller who would bring the country into his home; for the transient summer resident in the country; for the trail-erite; or for the Sunday drivers who perhaps have neither time nor space for a garden of cultivated flowers, there is a wealth of interest in the use of wayside plants as home decoration. And it is here, too, that we get best acquainted with the flowers. The use of wild flowers, such as the goldenrod, and the black-eyed Susan as bouquet materials is obvious, but actually, almost any plant or flower may be artistically used for decoration, if but a little thought is given to the matter.

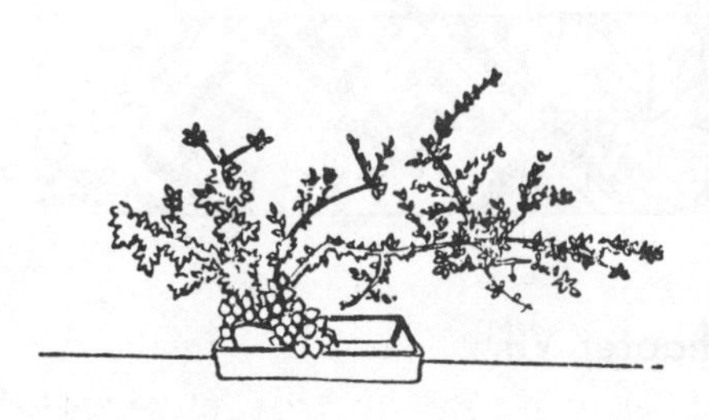

Flower arrangement itself, is such a personal matter, and also one requiring so much study to achieve the finest results, that it is quite impossible within the limits of this suggestive chapter, to propound any theory about flower or plant arrangement. The bibliography lists several books on this subject, and to those who are interested in meaningful arrangements, attention is called to the Japanese theories of flower arrangement. Certainly beautiful things themselves may be spoiled by faulty arrangement, and perhaps the best advice is to remember that there is a real beauty in simplicity and restraint. In advising the use of wild flowers for home decoration, one comes naturally to the matter from the conservationist viewpoint of the wisdom of gathering wild flowers. There are many rare forms of orchids and other flowers which should never be taken if we are to preserve the species for future generations. But on the other hand, there are many plants which may be gathered without loss to the world in any way. A few basic considerations should be noted.

1. Never pick orchids or other plants about which there seems to be a doubt. If there are only a few flowers growing of a kind, leave them in their own home.

2. Never pull up plants by the roots in order to gather them. This is especially true with such flowers as the trailing arbutus. In cutting branches from a tree, never destroy the symmetry of the tree. Random branches may be trimmed off with benefit to the tree, if done with discretion.

3. Never *break* branches from a tree. Cut them with a sharp knife or shears.

4. As suggested in our preface, be sure to obtain permission for cutting or digging any plants.

As every plant has personality, it may be helpful to add here some suggestions for the decorative use of plants throughout the seasons, and again it should be recalled that mention here is being made only of plants listed in this book. There are many other trees, flowers, and dried grasses which may be even more desirable, that you will want to use for decorative purposes.

SPRING USAGE

The *Wild Barberry* in blossom is a lovely yellow, and alone, or in combination with other spring flowers, makes a pleasing bouquet.

The knowledge of the beauty of *Dogwood* is too well known to require discussion, but it is well to note that however strong is the wish to cut this flower when one is out driving in the spring, it is not only a futile but often illegal act for the cut flowers do not keep well in water, and in any case are usually wilted before one gets them home. Dogwood should not be picked without permission under any circumstance.

Dandelion. Childhood's first gathered bouquet is apt to have been dandelions, and while the dandelion provides small scope for artistic expression, the sophisticated arranger will find that the seedheads are very effective and dainty, if they can be picked at the right stage. To use them in decoration, pick the stalks before the pods open when there is just a little white showing. Place these stems in a carton with the heads supported and the stems straight down. The stems will dry in a stiff condition and the heads will open so that they can be displayed, and yet the seeds will not easily be dispersed.

Plums and cherries. A mass of cream-white flowers is displayed on the black stems of the beach plum (as well as on other members of this family), while all of the flowers of the "stone" fruits make artistic arrangements quite without equal. The Chinese and Japanese make a festival on the flower-opening date of these trees, so beautiful do they consider them, and one artistic branch of plum may become a veritable shrine in the home.

Fern croziers. When out in the woods in the early spring, perhaps gathering the unfolding fern fronds for eating, watch out for some of the non-edible sorts of ferns to use for artistic arrangements. In the "dynamic spiral" of the young fern, one finds an art form which is often photographed, and which can lend itself to a short-lived but beautiful arrangement.

Skunk Cabbage. A bouquet from Skunk Cabbage? Yes, indeed. Cut the flower or spathe as it emerges, and arrange in a low bowl with a few of the bright green leaves of the plant, and you will find something not only beautiful, but wonderful to contemplate. It is not objectionably odorous at this flowering stage.

Mullein. Many professional flower "arrangers" use, in the spring, the soft woolly gray rosettes of mullein, as an artistic base for a larger flower arrangement. One of these giant rosettes when found growing in a rich farmyard pasture, is indeed a lovely thing, and the plant is so much of a weed that no one minds its being picked.

SUMMER USAGE

Asparagus. The soft feathery foliage of the wild asparagus plants, make a very useful "green" for combining with other flowers, and may well take the place of the florist's cultivated asparagus for that purpose.

Pearly Everlasting. This is, as the name indicates, an "everlasting" and the dried flowers are a good base for a dried bouquet. (If you like such things).

Water Lilies. The urge to gather water lilies is one that comes early to the country child, and yet is one that is just as pleasurable to the adult. The fragrant flowers may be combined effectively with the blue flowers of pickerel weed for a delightful and cooling summer bouquet. Note that in picking water lilies, they should be cut rather than pulled, as the pulling is apt to dislodge the roots from the mud, and spoil the plant for future growth.

Sumach. The red seed clusters of the sumach in summer, are often used with other materials to make strongly effective floral displays.

Narrow-leaved Dock. This is a universal weed and otherwise of little use during mid-summer, but the rich brown seedstalks of dock are particularly effective in arrangements for bouquets at any season. They should be harvested and dried by hanging upside down when they are fully developed and a rich brown in color.

Cattails. The dried heads of cattails are sort of obvious as a decoration. They combine well with grasses and other similar dried material. They should be cut (hung upside down) and dried in midsummer, before the seed is fully developed.

FALL USAGE

Milk Weed Pods. These pods are a prime source of winter decoration material, and are often found in florist shops in a painted condition. The inside of the pod may be painted in bright colors, and the outside sprayed if desired. The flowers themselves are pleasing in color, while the half-open pods are extremely artistic when suitably arranged.

Winter Berry. This deciduous holly is commonly collected in swamps, and sold along the roadsides everywhere in the territory where it grows, because the bright red berries can become the dominant red note of much Christmas decoration. The harvesting of the berried tops of these shrubs does no harm to the bush, and in fact, is apt to induce a greater growth the following year.

Juniper. A fine basis for winter bouquets are the berry-laden branches of the several native species of junipers. Not every tree has berries, and each is different in color, so that one should learn to be discriminating in order to secure the finest-colored and heaviest-fruited branches of the red cedars.

Maple. The well colored foliage of the sugar maple makes a short-lasting bouquet that is hard to resist when one is traveling through the New England countryside. It is the foliage of these sugar maples, the swamp maples and the oaks that makes the New England section the mecca of the whole United States and brings breathless gasps of wonder from foreign visitors. There is literally nothing to match our fall foliage display anywhere in the world. If it is done at the proper time and done carefully, the colored foliage may be dried and/or sprayed to preserve it for some time, for home arrangements.

Poke Berry. The purple-red foliage and fruit of pokeberry provides a lovely touch of color for fall arrangements in the home, especially when the bouquet can be put in front of a window where the sun can shine through the beautiful translucent leaves.

White Pine. Although this is not the traditional Christmas green, the foliage of the white pine is extremely soft and pliable and makes beautiful Christmas roping. It also provides a delightful pungency in the home.

Barberry. Just as in the spring the blossoms of the wild barberry provide a yellow bouquet, so in the fall, the gracefully drooping berries go well with fall decorations.

Japanese Honeysuckle. Because of its almost evergreen nature, the vines of the wild honeysuckle are quite useful for special table arrangements, where a low green is desired. This foliage may be used at any time of the year from early summer until winter.

Although the plant is not included in our regular list, the author would be remiss if he did not mention *Queen Anne's Lace* as one of the most important of wild flowers for home use as decoration. There is a graceful beaúty about this flower which makes it useful either by itself in a bouquet, or in connection with other flowers. The plant is a weed of the first water, and no one would object to its being picked. This is also true of many wild flowers like asters, goldenrod, daisies, Joe-pye-weed, and milfoil, which are found on every roadside.

While you are gathering flowers for use in the home, be on the lookout for natural "accessories" to combine with your arrangements. Often an interesting dried branch on a tree will have artistic angles, and may well become important in a low bowl arrangement. Also moss on lichen-covered stones may be useful in bringing an outdoor atmosphere to your arrangement. It is when you hunt for things like this that you will realize that beauty is to be had for the asking, and may easily be found along any roadside. It was one of the greatest of all artists who demonstrated this fact when he composed one of his great paintings by solely depicting a square foot of ground containing a patch of grass and a few wild flowers. This was Albrecht Dürer of the 15th Century. If you would see how to observe nature, you should look at this picture. It is when you take the wayside plants that are so beautiful in their own setting, into the home, that you will learn to appreciate the beauty that each has in itself, and learn, (as your skill as a flower arranger develops) to create an art form that is not at variance with nature for, as Sir Thomas Browne wrote:

> Nature is not at variance with Art, nor
> Art with Nature . . . Art is the perfection
> Of Nature . . . Nature has made one world
> And Art another.

WINTER USAGE

Except for the gathering of cut evergreens boughs for holiday decoration and the selection of mosses and little "groundlings" for the making of terrariums there is not much to gather in the dead of winter except enjoyment of the patterns of snow-covered branches and bright spots of berries in the landscape. But for the provident one there is the possibility of creating in the summer a device which will bring back quite literally a breath of summer: The preparation of a rose jar filled with pot-pourri. Although the majority of the ingredients fall outside the scope of this book, the fact of the use of rose petals will permit the inclusion here of a carefully tested recipe for POT-POURRI.

POT-POURRI

Although the necessary abundance of the petals of wild roses is a very doubtful circumstance, the rose garden enthusiast would very likely be willing to cooperate with you in providing, in rose season, an abundance of "just-past-their-best" roses that are ideal for drying as a base for some delightful pot-pourri. If you are interested in such a summer hobby, the paragraphs which follow will sum up the experience of some considerable research and experimenting which resulted some twenty years ago, in making batches of pot-pourri, which today are still delightfully fragrant.

If you are the fortunate owner of a lovely rose jar, which was purchased perhaps for its own ceramic beauty, you will find that your pleasure in the jar will be much greater if you can prepare for it your own supply of pot-pourri; or, again, the preparation of a supply of this fragrant mixture is a wonderful source for appreciated Christmas gifts, as the author has reason to know. And, furthermore, if making it for one's self, it is doubtful if pot-pourri would have to be made more than once in a lifetime, for rose jars will often remain fragrant over a period of three to four decades.

Some years ago in preparing pot-pourri to fill an inherited, beautiful, old jar, the writer made an analysis of many recipes which were found in herbals etc., and from these made up a fundamental recipe, which is herewith given for anyone interested in preparing the fragrant mixture.

First the dried rose petals. If, during the summer season, there is a supply of rose petals available from any kind of roses, the best time to gather them is late in the day when there is no dew or moisture on the flowers. The next day the petals are to be spread out in the sun on a tray, making sure they are well out of reach of a gust of wind or a possible shower. Now and then the petals may be shuffled a bit to turn them over. This drying process is to be repeated daily when the weather permits until the petals are properly dried. In this condition they should not be crisp or dark brown, but just so that, while they still retain a certain rose-petal color, you can feel that there is no moisture in them. If one batch of petals is dry and others are still to come, put the dry petals in a covered dish and mix them with a pound of salt. Open the jar daily and stir the petals and salt. The salt is added in order to take out any excess moisture, and the daily stirring prevents molding. After several weeks of this drying process, you will have sufficient dried petals to proceed to make your pot-pourri mixture.

If the season or your garden conditions have not been right for the production of roses, you should not give up trying to fill your rose jar. Dried rose petals are imported into the United States from northern Africa and may be purchased in most large cities. Such commercially dried petals will answer every purpose except the sentimental one of producing one's own supply. (One such importer of dried petals is the Cheney Drug Co., 14 North Street, Boston, Mass.).

During the drying of the rose petals, you will want also to be drying separately, sprigs of certain of your favorite fragrant herbs. In the case of a plant like the rose geranium, it takes a long time to dry the leaves because of their high water content. In deciding which plants from your herb garden to use in the rose jar, remember that really pungent plants like lemon verbena should be used very sparingly so as not to make one fragrance more noticeable than the rest.

When you are ready to assemble your ingredients, you should provide yourself from the drug store with a quantity of ground orris root. For two quarts of dried rose petals you will want two ounces of the ground orris root. In addition (and for the same amount of petals), you will want one ounce of gum benzoin and four drops of real Attar of Roses, if it is available or if the genuine article cannot be obtained, you can use instead a little larger quantity of synthetic rose perfume.

From the corner grocery store you will need in addition to the salt already recommended: Three ounces of brown sugar, an ounce of brandy or pure alcohol, one ounce of whole cloves, two ounces of allspice, and one ounce of stick cinnamon. These spices are to be ground at home with mortar and pestle just before the pot-pourri is assembled.

When you have secured all the above items, and you are assured that your petals are completely dried, your mixture is ready for assembling. The first thing to do now will be to sift out the bulk of the salt from the dry petals, and mix together in a large bowl the dried materials, the sugar, spices, benzoin, orris, and a few dried sprays of the several herbs which you have selected as being desirable. This mixture is now to be set aside in a covered jar for a period of three or four weeks, during which time the contents should be stirred *once a day* so that the various oils will fuse together into one grand perfume. This daily duty is a most delightful part of making pot-pourri because of the wealth of fragrance that greets you as you open the jar to stir it.

When you are satisfied that the stirring process has produced an infusion of all the odors and that the material is thoroughly dried, you can fill your rose jar, adding at the end two or three drops of the Attar of Roses in each jar, and stirring or shaking it together. If the mixture seems to be too dry, you can add a little brandy, which, with the additional rose essence, is the procedure to follow when, after four or five years, the rose smell has gone.

If the recipe which we have recommended above has been followed in its entirety, you will find that you have a considerable quantity of this mixture, perhaps more than will be taken by one large jar. If you have a surplus, by all means keep it for use at Christmas time. Pack it in neat cellophane bags tied with ribbons and send to your friends or use it to fill the miniature rose jars that one will often find displayed in the stores at the holiday season.

A month or two after this mixture is made you may perhaps find that the smell of cinnamon or other spices is a bit predominant, but time will smooth all of this out into one blended odor that will be oriental and delightful. By midwinter in any case you will find that you have a constant reminder of the joy of summer, which will bring before you the scenes of a summer rose garden, and which will certainly provide a "conversation piece" as stimulating to the nose as to the memory.

PART II — *Plants Outdoors*

NATIVE PLANTS FOR LANDSCAPE USE

The history of gardening shows us that for many thousands of years, man has gone into the fields and taken into captivity the plants that grew there, to use for his own pleasure. At least, one can say, that his garden started in that way, but as it became fashionable to use things from distant places, man soon came to prefer the "imported" plants, as having more prestige value if not more beauty. One can well imagine that the King of Babylon required plants for his Hanging Gardens to come from the surrounding countries, just as we know that early Japanese gardeners brought to their own country for domestication, all of the finest plants of China. So, too, today, do we in America seem to prefer exotic plants in our gardens, neglecting many beautiful things which are available in the woods around us. It is true that the use made of wild or native plants does not always provide the greatest challenge to our cultivating skill, nor sometimes are the flowers the largest, but increasingly governmental agencies, if not the citizens, are realizing the value in native plants for landscape purposes. One notices that the great toll roads, the freeways, and the principal highways are being beautified with trees and flowering shrubs indigenous to the section through which they pass. Outstanding in that connection has been the use along the Connecticut parkways of vast quantities of the native laurel and dogwood, plants that have no equal in floral effect among all the plants of China. Shrubs such as these are naturally perfectly hardy in their native clime, and home owners may well take a lesson from this carefully considered landscaping along the modern highways.

Not only can one use trees and shrubs in this way, but also a selection of the lower growing herbs of the woods. Some of these not only are beautiful but provide a definite challenge to the proud horticulturist. There is no garden flower from whatever country that can equal the brilliance and beauty of the native Butterfly Weed, (Asclepias tuberosa) but anyone who has tried it, knows that it is well nigh impossible to domesticate. The same may be said of the all too rare Cardinal Flower, (Lobelia cardinalis) which is ever happiest in its own native habitat in the roadside ditch.

There are then, many plants which can be transplanted from among the wild growth of the roadside to the more formal surroundings of a home landscape. It is impossible to delineate all of these plants, but within the prescribed limits of this book, one finds many which happily can be used for domestication.

First a few words about the trees which may be transplanted. Among suitable ornamental trees, first mention would go to the Sugar Maple, which is among the finest of all shade trees and grows easily quite without pests. Second in choice among wild trees which can be used in the home grounds are the Oaks, good, but difficult to transplant. The several species of native Junipers mentioned are useful for transplanting if carefully handled and selected, while the various nut trees are all suitable for use if the home grounds are spacious enough. Rarely can the small home property accomodate the eventual great size of the White Pine, while among the Birches the death rate is so great as to discourage their use.

The principal difficulty in using trees from the wild is that they are not easily transplantable. If you are in a hurry for results, a nurseryman is best consulted, either for the actual digging and transplanting from the woods or for the purchase of a nurserygrown tree. If, however, you wish to do your own transplanting of wild trees, the venture should be planned well in advance. If it is possible to select a suitable maple or oak tree in the wild, pick one that is not over three inches in diameter at the base. Anything larger than this would be almost impossible to move, where it has not had the root pruning of a nursery grown tree. If the plant can be moved in the winter time with a ball by professional movers, perhaps a larger tree than indicated can be used and taken directly from the wild, but if it is to be a home venture (or a one man project) it would be best to dig around the tree for a distance of 24 inches, about a year previous to the time of transplanting. This will cut the main feeding roots and allow the formation of new root growth, thus decreasing the shock of moving. The same procedure should be followed in the case of trees of any kind, especially the evergreen ones, because with root growth going widely in search of food and water, it is hard to move them from their original home with any degree of success. The proper season for moving most trees would be in the very early spring, excepting again evergreens, which may best be transplanted in late July or early August.

Among the *Shrubs* of our region which are suitable for landscape effect, one might mention the following:

Sumach	6'	Common Barberry	7'
Japanese rose	3'	Sweet Pepper Bush	6'
Elderberry	7'	Deciduous Holly	8'
Witherod	6'	Bayberry	3'
High Bush Cranberry	7'	Beach Plum	4'
Blueberry	4'-5'		

The approximate ultimate height of this group of our native shrubs is given, following each name. It is important to remember these heights when planting about the home, because when plants are small it often seems impossible that they can grow so tall, but under normal growing conditions, they do grow at least to the heights as indicated.

All of the above shrubs do extremely well under cultivation, with the possible exception of plants like the beach plum and bayberry, which seem to especially be happy only under seashore conditions. The sweet pepper bush and the deciduous holly are most at home in a wet place, but will grow reasonably well under home garden conditions. Almost all of those mentioned seem to do best in a slightly acid soil, and respond well to fertilization. This statement is especially true of the various forms of blueberry bushes, which usually require special treatment or woods soil to provide an acid soil condition, but given this and with proper fertilization the blueberry makes one of the most beautiful of ornamental shrubs. Certainly the brilliant and long lasting foliage of the blueberry in the fall alone makes it worthy of more frequent use as a landscape plant.

Among the *Vines* which one might mention from our list, are the Japanese honeysuckle and the wild grape. Both of these vines have their place in the home landscape picture. The honeysuckle is especially valuable as a ground cover or a vine to prevent erosion on a sandy bank, and in such a spot it will grow unattended for many years. The wild grape on the other hand, makes a wonderful trellis subject in addition to its production of fruit.

On the subject of Ground Covers, perhaps one of the best ground covers in existence for seashore sections is that of the *Bearberry.* When one sees this growing in the wild, covering many square yards of ground with its beautiful shiny green leaves set off by bright red berries, it is hard to realize that anything can be more beautiful. Many people, therefore, have tried to transplant this plant to their homes, with no success. Frankly, no plant is more difficult of handling than the Bearberry and even nurserymen shy from it's culture. A recommended method of procedure in transplanting this plant, is to take a four-inch clay pot and sink it in the ground (filled with sand) alongside of a naturally growing plant of Bearberry. Some of the little vines can be placed on the top of the pot, where they will root naturally into the pot. Later in the season or perhaps the next year the vines from the mother plant can be cut loose, giving a potted plant which may then be taken up and transplanted to a new spot which has been prepared to approximate the natural sandy condition. This seems like a difficult procedure, but it is no worse than losing all the plants transplanted, and the final result is well worth the effort.

For those who would have a pool on their home grounds, there are several Water Plants in this list which would be happy in such a damp situation. We include here plants like the Cattails and the Iris, and Sweet Flag, and, of course, in the pool itself, some of the native Water Lilies, which have no equal for fragrance.

In the list of herbaceous material or flowering plants, there is not such a wide selection of suitable garden plants. Perhaps ferns take the first place among plants which may be transplanted from the wild to the environs of one's own yard. The Milkweeds are also desirable to use in this way, as previously mentioned. The Mullein, weed that it is, would be suitable if one had a large enough garden in which to accommodate its rank growth. Of course, the Mints would be useful additions to any garden, for their fragrance. To use in a mixed shrubbery the poke-weed could be well recommended, because of its beautiful fall coloring. Goldenrod, too, although not normally used in American gardens, is actually a wonderful border plant and is so considered among the Europeans.

The above will be enough to show that there is a wide variety of wild material which can be used in any American landscape planting, and these notes are written to urge the consideration of such use. For instructions for the design of landscape plantings, the attention of the reader is directed to books on the subject which, unfortunately, usually refer only to material purchased from nurseries. The writer, for many years a nurseryman, appreciates the desirability of using nursery-grown material, but realizes that there are times and places when either the desire to use native material, or economy, dictate the beautification of one's home yard with some of the bounty of the wayside.

SECTION 2

Notes On The Plants

In the chapters that here follow, are described and illustrated the various 100 plants whose uses have been detailed in the first seven chapters. Descriptions are deliberately brief and the illustrations have been selected to present some special characteristic of the plant or its usable part, rather than to offer photographic realism. Included for most plants will be found a map of the Eastern United States showing roughly the distribution by states of the particular species. This device, based on the 8th edition of Gray's Manual Of Botany, must be regarded as an attempt to show you where to look for the plant, but only in a very general way, for beyond regional habitat must go a consideration of the usual home of the plant — elevation, soil, and other factors.

Again — the inclusion of various trees, shrubs, etc. in classified chapters is a decision based on a personal selection, and with these choices some readers may disagree. Botanical classification and nomenclature is ever in a state of flux and reflects the minds of many people, and therefore no apology is offered for the author's personal distribution of the plants into the following five chapters.

It is simply felt that here we have the 100 or so wayside plants that every child, every nature-lover, and car-owner should know about and that essentially the arrangement by chapters is unimportant.

In the Index, which is the most important part of this book, will be found the page number for each plant and for the pages where its use is elaborated.

Chapter VIII
LOOK UP — USEFUL TREES

"One impulse from a vernal wood
May teach you more of man,
Of moral evil and of good,
Than all the sages can."

WORDSWORTH — "Tables Turned"

Acer saccharum — Sugar Maple

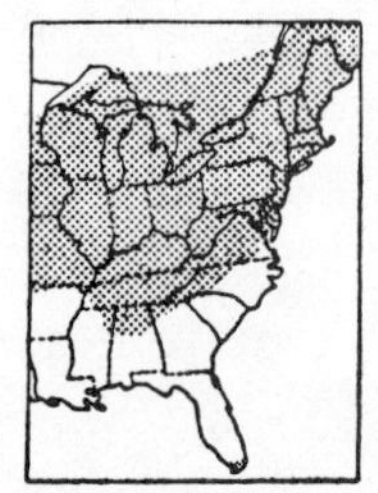

STANDING ALPHABETICALLY AT THE TOP of a list of trees is the Latin name for a native, which, in the opinion of many, is at the top of any list of *valuable* trees in the eastern United States. It can usually be identified by the gray and deeply furrowed bark of the older trees, the characteristic shape of the maple leaf, and the double-winged "samaras" or seeds. It is from Acer saccharum that our maple sugar comes, but it should be noted that other forms of maple also have sap of a variable sweetness. Both the sycamore maple and the box-elder contain considerable sugar, while in a pinch, one could secure sap from such trees as the hickory, the birch and the ash.

Without question, the use of maple sugar came to us from the Indians, who would boil it by throwing red hot stones into buckets of sap until the steam had reduced it to sugar. But we in New England know that the shade of the maple and its clean habit and fine fall color make it not only the source for the prosperity of Vermont but the best of all shade trees, especially now that the elm is being attacked by disease. The wood of the maples is also extremely valuable, because it is hard and workable, and in light-colored furniture is much desired for modern interiors. Again the bark of the maple tree was used as a brown dye in the early days, and (even in old age when the tree must be cut down), it provides the finest of firewood.

In the recipe section, various uses for maple sugar are given, but at this point a few words about sap gathering may be helpful. The actual date for tapping a maple tree is hard to give, but with winters getting warmer, as they seem to be, the sap may well start to rise after the bitter cold weather of January is over. It should be remembered that the problem of getting sap is to have alternating, nights of freezing, and days of thawing.

To tap a maple tree, drill into the south side of the tree at a convenient height, a hole 3/8 to 5/8 of an inch in diameter to a depth varying from 2 to 4 inches, according to the thickness of the outer bark, and with the hole slanting slightly upwards. Drive into this a metal sap spigot or, failing to secure this, prepare one from a thick piece of elderberry bush stem, from which the pith has been pushed out. In any case, one should have the hole a tight fit for whatever spigot is used, and it should not be driven into the full depth of the drilling. Provide now a suitable bucket for the sap to run into, and, if possible, cover to keep out the dirt and flies, or diluting rain or snow. Much more sap is given than one would expect on bright days, so do not forget to empty the bucket daily. On very large trees, one or two taps may be made on one tree, and this will provide enough maple syrup for the average family.

The boiling down is now the problem, and where only a home stove is available, the sap from one tree is all that can, with limited facilities, be boiled down, as several days will probably be taken in the process. There is no trick to doing this except that one simply has to boil and boil until enough water has evaporated to leave a syrup of the desired consistency, which should, before bottling, be strained through flannel. Or one can carry the boiling to the point of crystallization of the syrup into maple sugar.

However much the expense or trouble, the pleasures of tapping a maple tree are very great, and there is no thrill quite as fine as having next winter's pancakes covered with your own home-made maple syrup. Nor is there any drink so satisfying as a drink of sweetish cold sap direct from nature's soda fountain.

Betula — The Birches

Throughout all the northern or mountainous areas of the northeastern states, the various species of the birch are perhaps the most noticeable of all trees, standing out as they do, with their generally white bark against the rich green of the conifers. A number of the birches are useful in various ways, the species worthy of mention here being:

Betula alba: Canoe birch, Lady birch, Paper birch, White birch; *Betula lenta*: Black birch, Cherry birch, Mahogany birch, Sweet birch; *Betula lutea*: Yellow birch.

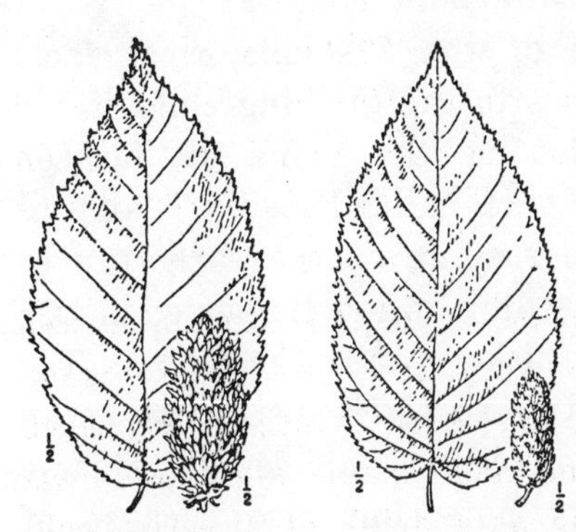

Yellow birch. Sweet birch.

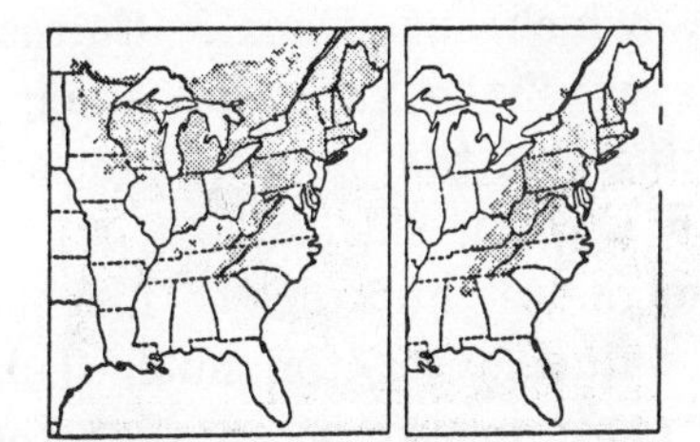

One of the commonest uses of the birches is for beautiful, if not efficient, fireplace logs in city homes. The bark has been eaten in northern countries in times of famine. The sweetish sap of the white birch in the past was sometimes collected and made into vinegar. It is interesting to note in the case of the sweet birch, that the oil extracted from this tree is sold commercially as *oil of wintergreen.* This same extracted oil of birch also was the principal ingredient of real home-made birch beer. Sweet birch is a desirable source for a cooling drink or tea, because of flavor and a natural sweetness.

Again, from the fresh or dried leaves of the birch a good yellow dye is obtained.

Actually we see birch bark most commonly used for the making of novelties for gift shop sales and from this useful bark the Indians made their canoes and in one way or another the birches had a practical and even vital value in their lives.

For the average highway traveler today the beauty of the birches is often quite enough, and is a worthy subject for photography and appreciation.

Carya ovata — SHAGBARK HICKORY

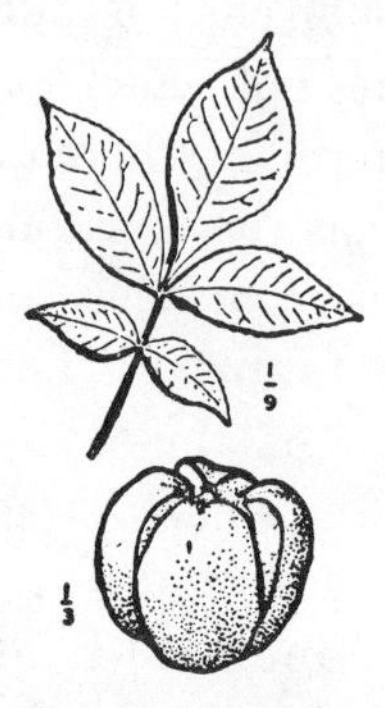

Shagbark hickory.

One of the first historical facts that comes to the growing child is the association of the name of Jackson with that of "Old Hickory," and as well, the reading of the early life of Lincoln, telling of how he earned money splitting hickory rails for fences. The hickory tree surely is distinctive as an American tree, not only for its valuable and tasty nuts, but also because of the unusual character of the bark of the tree, which, as Donald Culross Peattie, says, "is a rude and shaggy coat, looking like the pioneer, himself, in fringed hunting shirt."

Although we think of the hickory as being valuable for the nuts, it is probably more prized because of the special nature of the wood of the tree. Toughness is the first of these qualities, and there are many uses for such an enduring material. Our early American fences were often made of split hickory, as were many parts of wagons and the handles of tools. Also in the days before coal was commonly used, the hickory was the prime fuel of the early settlers, modern statistics now showing that a cord of hickory is almost equivalent to a ton of hard coal.

Last, but by no means least, is the value of the green wood of the hickory for smoking hams and bacons. It is this use which keeps us all more or less in constant touch with this important tree. Unfortunately with the advance of civilization, the hickory tree is gradually disappearing from our woods, the large ones long since having been

CHECKERBERRY HICKORY NUT

cut and the new ones finding it harder to get established. No one, however, who has ever gone hunting in the fall for hickory nuts, can deny the many attachments that one can have to the hickory tree, nor its place near the top of the list of valuable American trees.

THE WALNUTS

Juglans nigra — BLACK WALNUT

Although possibly, originally not native to most parts of New England, the black walnut, through plantings, has spread to all but the coldest regions of the northeastern states, and to other sections of the country. So widespread was it that it was one of the most important trees of the 19th Century, and this largely because of its use for lumber rather than for the tasty nuts. The black walnut can easily be identified by the typical compound aromatic leaves, as well as the rather thick hairy husk which protects the shell of the nut. The black walnut is one of the great trees of this area, often in the past reaching a height of 150 feet, with a straight heavy trunk, but through the years the large ones have mostly been cut.

Certainly the black walnut is one of our finest native nut trees, having a kernel of a quality and flavor, quite distinctive. The only difficulty being that they must be harvested at just the right time and under just the right conditions, else the oil will turn rancid, and the fruit be ruined.

As indicated in the chapters on recipes and dyes, there are many uses to be made of the fruit. The stain of black walnut shells is extremely powerful, and may be used for many purposes. Fernald in his book indicates that not only is the dye good for cloth, but also makes beautiful stain for leather or wood. For this later purpose he recommends adding a little benzoate of soda to the extracted juice of the hulls, to keep it from spoiling.

As a cabinet wood, the black walnut with its oily nature and dark color takes a very high polish. Long out of fashion, the use of walnut in furniture is now being revived, but as is true with so many other trees, the supply of the best timber is vanishing from our woods. The joy, however, of gathering this or any other sort of nuts in the fall, should be one of the rights of childhood, as well as a productive privilege for the adult. For all these reasons everyone should learn to recognize the black walnut tree.

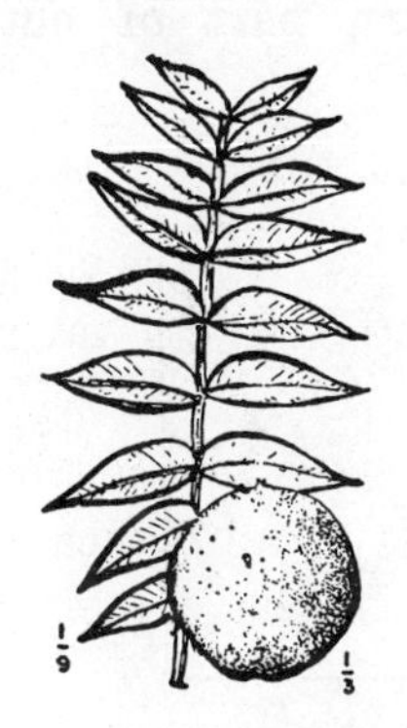

Black walnut.

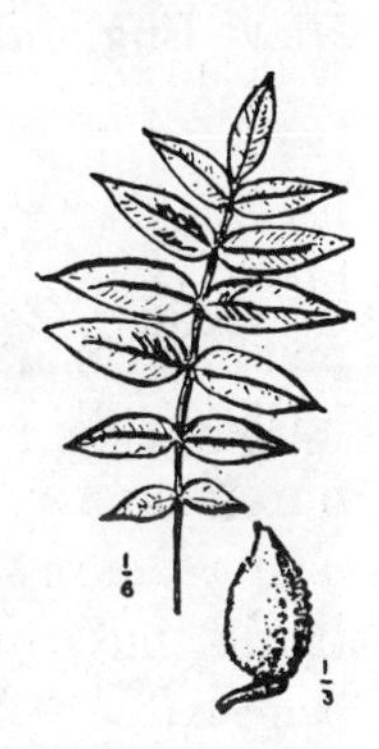

Butternut.

Juglans cinerea — BUTTERNUT

The butternut tree is very closely associated with the black walnut, and can be distinguished from it somewhat by the difference in shape of the compound leaf, and the rather elongated appearance of the outer husk of the nut itself. The fruit is much less distinctive in taste, but where found, is extremely desirable for use. The cabinet wood made from the butternut tree is much lighter in color than that of the black walnut, and was much used in early days for furniture making.

The Shakers, that interesting communal religious sect of the 19th Century, used the butternut tree not only for furniture but for the production of a rich purple dye. Other early settlers found that an extraction of the inner bark of the root was a cathartic, but such a use today is not so important.

The butternut trees will be found to bear at a much younger age than the larger black walnut. The fruit is a little harder to shell than is the case with some other nuts, but their mild flavor make butternuts ideal for use in candy or cakes. One will find the butternut tree growing in rich soil almost anywhere from southern Canada, across New England and down to the southern part of our territory.

Corylus americana — THE HAZELNUT

In our region there are two species (*C. americana* and *C. cornuta*) of the hazelnut but the hunter for nuts in the fall must not expect to find the same grade of hazels as one would of butternuts or hickories. The hazelnuts in our country are all to apt to be wormy but a few can be cracked that are fit to eat. The flavor of good ones is fine and the food value high. The illustration well shows the typical leafy husk around the nut, which is the identifying sign of the hazel; that and the rather coarse foliage which has an appearance of birch leaf. Some might class the hazel as a tree but generally it is found growing as a large shrub, often on sunny roadsides.

Cornus florida — Dogwood

Perhaps one of the most beautiful spring flowers in this part of the country is the dogwood, which is always welcomed when it whitens the woods and roadsides early in the season. In the main there is but little use for the Dogwood for food or medicines, but it is worthy of mention in a book of this nature even though it is too well known to require description or identification.

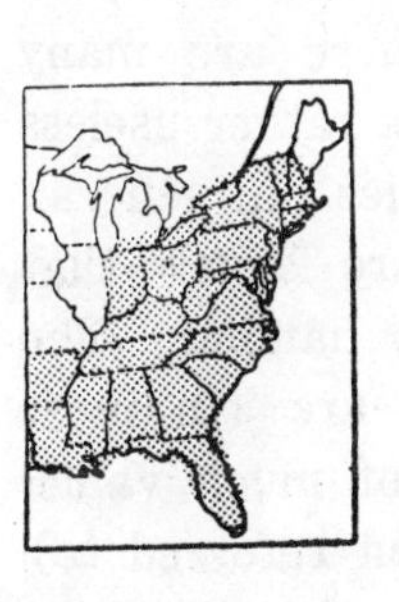

At one time it was thought that the bark of the dogwood contained medical properties similar to that of quinine, but such properties, if present, are of little value today. The extremely hard wood of dogwood has often been used for the making of tool handles, while its grain and color make it ideal for ornamental cabinet work. One possible further use of the tree which has some historic precedent, is the use of the young branches, stripped of their bark, for making toothbrushes. For this purpose many Indian tribes so used this tree, and found additionally that an application of the juice of the twigs was useful in preserving the gums. Dogwood flowers, when cut, have little if any value for decoration, since they wilt very rapidly. (The dogwood tree is the state tree of Virginia.)

Crataegus species — HAWTHORN

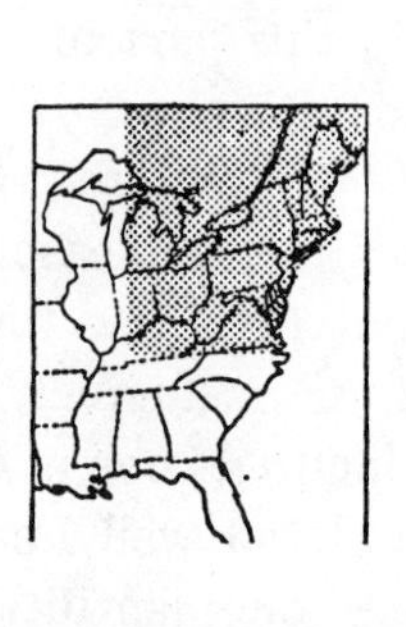

The Hawthorn tree can usually be identified by its small white, rose-like flowers and leaves, and the often large sharp thorns which protect it. There are many natural hybrids of hawthorn and hence it is rather useless here to identify particular species or varieties. In general, it may be said that the uses of the tree are largely that of providing material for canes and tool handles. The berries, while a possible source of food, are (in most species) rather too dry and pulpy to be of much value. The English hawthorn (the "May" so often referred to) provides berries succulent enough to make "haw" jelly.

Juniperus virginiana — RED CEDAR

Closely associated throughout most of our territory with the spreading ground juniper is the often tall and stately red cedar. This is one of the very desirable plants of the eastern United States. The berries may be used (as in the species *Juniperus communis*) as an ingredient for flavoring gin, and there are some records that medicinal use has been made of them as well. Both the bark and the berries provide material for a khaki-colored dye.

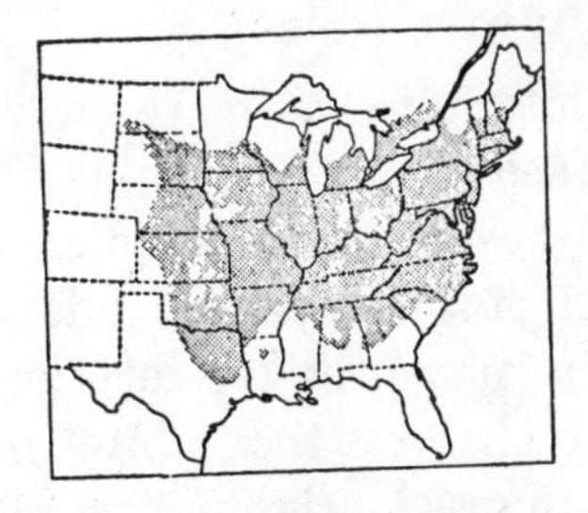

The richly colored heart-wood when shaved up into small pieces provides through its volatile oil, one of the best moth deterents known. Because also, of this high oil content of the heart-wood, pegs for the planking of ships are most desirably fashioned from red cedar.

In sections where the tree grows, the red cedar is often moved in from the wild as a tree for landscape planting about the home, but when using it in this manner, care should be taken to see that it is provided with a sunny spot. Cedar moves fairly easily from its childhood home if the moving is done in midsummer, care being taken to gather into a root-ball every possible bit of roots and all with the least delay. In selecting cedars for transplanting particular attention should be paid to the individual habit of the trees. The red cedar is an extremely variable tree, each seedling tree having characteristics as to shape, denseness of foliage, and color; the color running from a light green through almost pure blue, to (in winter) quite rich reddish greens.

The red cedar is much used as a Christmas tree in some areas. The writer, at home in New York State as a boy, never knew a fir tree as a Christmas tree, but only the red cedar, and that this use is still common, is shown by the fact that the Department of Agriculture statistics for 1952, showed that 10% of all Christmas trees used in the United States were the eastern red cedar.

Pinus strobus — WHITE PINE

Eastern white pine.

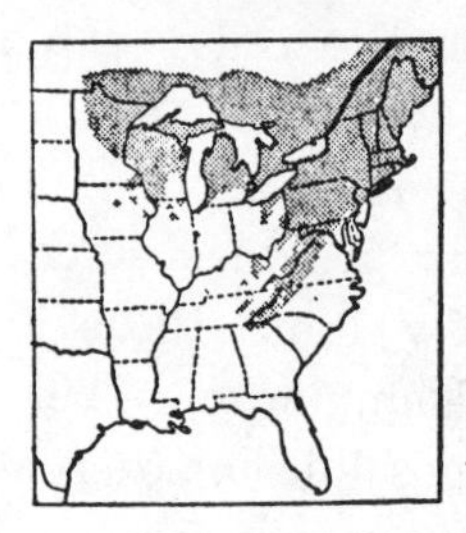

The white pine is more or less easily identified among other pine trees by the five needles in a bundle, and the soft feel of the foliage. It is one of our very noble native evergreen trees, which before the white man came, grew to great height. The white pine was indeed so important to the British in Colonial days, that all ancient straight white pine trees were called "mast trees." Before the Revolution, Emissaries of His Majesty, went through the woods to blaze them with the king's Broad Arrow, to mark them for later use as main masts for "ships of the line." Today, trees of such dimensions are rarely seen in the woods, but the white pine is still distributed widely and should, with its many fine qualities, be known to all.

First, is its value as a soft, workable, durable beautiful wood. One has but to see the many 18th Century houses built of pine, in and out, and still surviving, to appreciate how fine a wood pine is. But there are side values to the pine in the realm of food and medicine.

The cambium, or inner bark, of the white pine has always been known as an emergency food, even in early times as a source of material for candy by New Englanders. In Europe, northern peoples very often cut this inner bark in spring, dry and store it, to be used as necessity demands. Ground up, it was also mixed with flour to extend bread in time of famine. Linnaeus gives specific instructions for using it in this way, and we know that the American Indians also used it extensively. The seeds of the white pine are pungently spicy and were gathered by many Indian tribes for cooking with meat. Medicinally, the bark is a source of an ingredient in many cough syrups of today, one of the most commonly known ones being White Pine, honey, and tar.

Finally one should mention the value of the branches of white pine for Christmas wreaths and roping, for which it is ideal. Unlike hemlock and fir, white pine needles are persistent under our hot, dry, home conditions as the pine oil volatilizes and it brings to the home the true smell of the woods. One can experience the same pleasant odor in driving through pine country on a hot summer day and this pleasure should remind us that the white pine is one of our best native evergreens.

Populus species — THE POPLARS

There are a number of species of poplars largely to be seen in the northern part of the eastern United States, for the poplar is essentially a tree of the Northland. Possibly its present territory will be increasingly restricted as the climate grows warmer. In general, not too much good can be said of the poplars, beyond the fact that they provide material for the manufacture of paper; a discussion of which process is beyond the limits of this book.

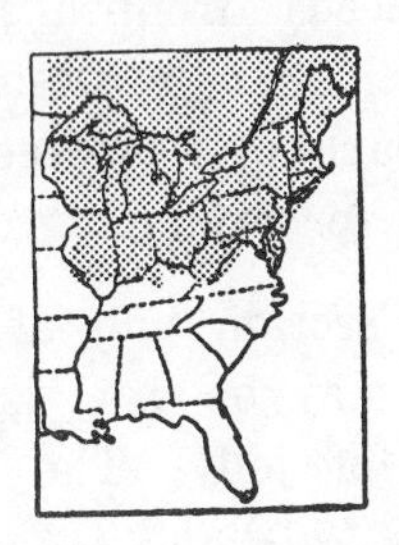

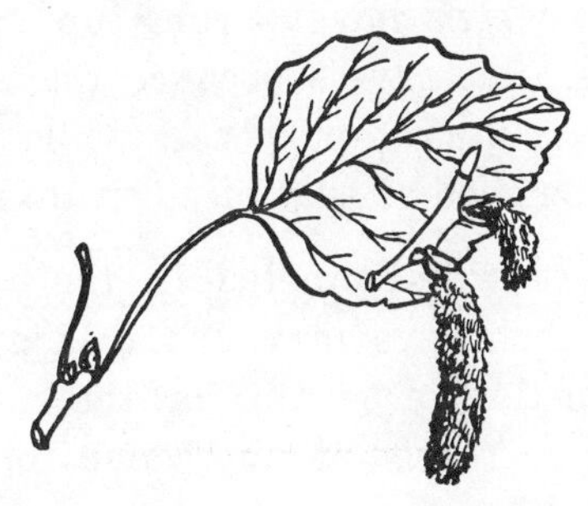

About the only good thing that may be said of most poplar species, is the fact that they will grow in rather poor soil and even seem happy under city conditions, but this still does not make them a good street tree for their soft useless wood and huge widespread root systems, both make for danger from falling limbs and clogged drains. Let us, for the most part, enjoy the poplar as it comes to us as the prepared pages of our daily paper.

The poplar is included here because the large fat buds of certain species carefully dried, may be used (in mixture with other herbal materials) medicinally as a cough syrup. One species known commonly as Balm of Gilead is a component of an unguent for sprains. Further, Fernald and Kinsey, in their book on edible plants, indicate that the fiber just under the bark called the bast, may be used as an emergency food.

Malus species — The Apple

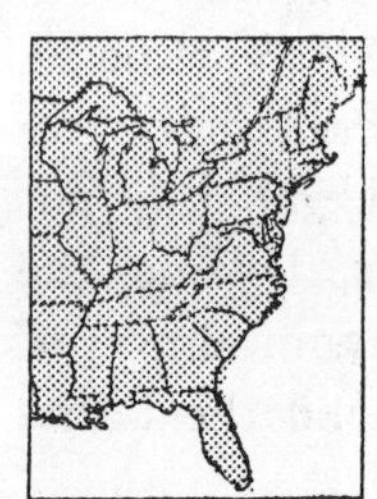

Johnny Appleseed was not the only agent by which apple trees became common in the United States, for nature herself has been very generous in her planting and distribution of wild apples. Everywhere along the roadsides in the eastern United States, one will find apple trees which have sprung up from seeds dropped from birds. Such trees produce fruits which may be used for many culinary purposes, while the bark of the trees is just as good as that of cultivated trees for producing dyes.

As will be noted in the recipe section one of the principal ways in which wild apples may be used is the production of pectin for the making of jelly. The wild fruit is rarely of any value to eat out-of-hand, and for

APPLE BLOSSOM

this reason there should be no complaint about picking such apples found growing by the roadside. Native *crab-apples* (as well as escaped sweet seedlings) may be found growing south and west of New York.

In summer, the wild apple is one of our very useful wayside plants, in addition to which there is its beauty and decorative value in the springtime. Delicate pink blossoms from the wild apple tree against the dark artistic framework of its branches can provide a basis for a flower arrangement second to none.

In discussing the use of wild apples (including the crab apples) it should be noted that many of the cultivated Oriental forms of crab-apples which the nurseryman often recommends and which are planted purely for ornamental use are also excellent for the making of jelly. A supremely beautiful jelly may be made from the cherry-like fruit of the red-leafed Polish crab, almost any of the others are also quite suitable for this purpose.

Much has been written about the use of apples for purposes other than cooking and eating, especially as regards the therapeutic values. Beyond this "apple-a-day" value, in both England and Germany pieces of apple have long been used as a remedy for warts. The use of cooked apples with pork or the addition of cheese to apple pie both are culinary customs undoubtedly with medical reasons. Our grandmothers knew a thing or two when they put slices of apple in the cooky jar to keep soft cookies, soft; while the modern cook knows that apple parings boiled in aluminum pans will clean them brightly.

Prunus species — WILD CHERRIES

There are two principal species of the wild cherry, *Prunus serotina* and *P. virginiana.* Both species are common to our territory, and one (at least) is a most desirable food plant. This *P. serotina,* is the one usually known as the black or rum cherry, whereas *P. virginiana* is known as the chokecherry. The differences between these two are largely in the arrangement of the clusters of berries. The more desirable rum cherry has fruit in long grape-like clusters of purple-black, the leaves having rather small blunt teeth, whereas the chokecherry has shorter clusters of fruit, quite strongly acid, the leaves having sharper teeth on the margin.

The chokecherry evidently earned its name because of the puckery quality of the fruit, yet the very same fruit when prepared as jelly, (as detailed in the recipe section) is most delicious, and there are many references to the use of these fruits as a flavoring for brandy. The

rum cherry which has a slightly bitter but very rich flavor, is even more ideal for making a jelly of a most beautiful color and a flavor which goes well with almost any red meat. Fruit-laden trees of both of these species are very common along the roadsides in the fall, so that there is no trouble in gathering them, unless the birds should have found the tree previously. It is the observation of the writer that it pays to notice the variance in the quality of the fruit of the wild cherries. Some of the trees, especially the young ones, will have very fine large fruit, whereas a neighboring tree may have fruit which is all pit.

Beyond using the wild cherries for jelly or wine one should note that they have considerable medicinal value. It is from the bark of the wild cherries that we get the extract which is so often the ingredient of good cough syrups. This is probably because of the prussic acid which is found in the cherries, which is said to be a sedative to the nerves.

From the fruit and the roots of the wild cherries comes a good reddish-purple dye. If trees of sufficient size are found, they will provide a beautifully-colored cabinet wood.

Quercus species — THE OAKS

Here is a family of trees about which a book could be and has been, written. Rutherford Platt, the nature writer whose books everyone should read, says that there are more kinds of oak than of any other tree. Our American landscape is filled with a great variety of oaks, many forms of which are ideal when transplanted to the home landscape. The oak tree has, indeed, through the ages made itself the symbol for sturdiness and few woods are so prized for various kinds of carpentry, as one or another species of oak. The longevity of oaks has also given us such famed trees as the Charter Oak of Connecticut, the Wye Oak of Maryland, the Treaty Oak of the District of Columbia, the Marshall Oak of Michigan, Waverly Oaks of the Boston area, and many another, notable for size, age or historical association.

Some of the Oaks are members of the family of white oaks and others of black oaks, each family having many interesting variations in leaf and growth habit.

In "usefulness," at least in the terms considered in this book, the oak is not high, its principal values being in the tannic acid found in its bark and leaves, the uses of such acid being discussed in appropriate chapters.

Salix species — THE WILLOWS

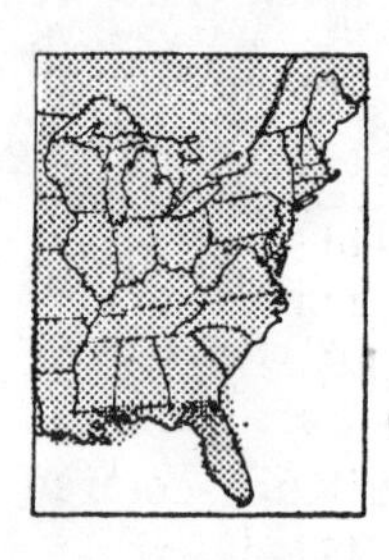

There is not much to say about the willows, except that the family is a very large one, with one species or another to be found growing almost everywhere in this country. The willow as a family is included here, because of the general usefulness of the pliable branches of some of the forms for use in basketry. The

wood itself is very hard to work, and the growing trees or shrubs have little value generally, other than for landscape planting purposes, and even here their uses are somewhat limited due to the great need of the willow for wet feet. The black willow (Salix nigra) is one of the commonest of the many forms. In many of the Colonial States the willow is best known, through the monumental presentation of the weeping willow (a native of China) on thousands of the 19th Century tombstones.

Tilia americana — BASSWOOD

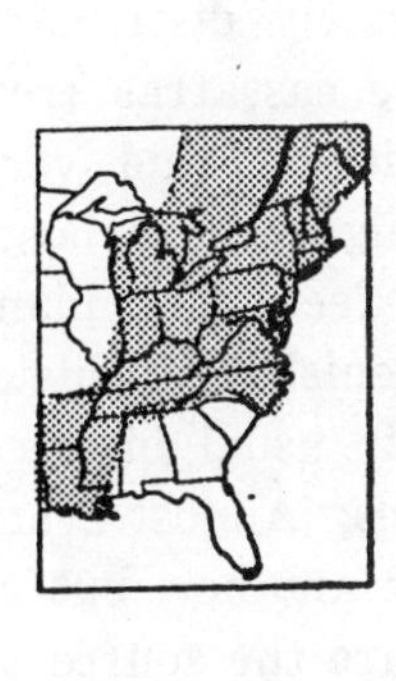

Although not quite as valuable an ornamental tree as the European form of basswood known as the linden, the American basswood is one of our largest eastern trees. It is valuable largely for the mildly fragrant flowers which provide a fine source of honey for the bees.

The American Indians, on the other hand, found the basswood a favorable source of material for making rope and twine. They separated the inner layer of the bark into slender strands and wove this into a sturdy rope. In Europe other species of Tilia have for centuries been used as a chief source of rope material. When the Basswood was more plentiful in early American days, the wood was much used for cabinet work where a soft easily-workable material was required, such as wood for carving.

Europeans generally, and the French especially, use the fragrant dried flowers for an interesting and possibly healthful "Tisane" or tea. It is the same European Linden, neatly clipped, which gave its name to the Berlin street "Unter den Linden" as well as many another like avenue throughout Europe.

Sassafras albidum — SASSAFRAS

One of the most interesting and widely distributed useful plants in the United States is the sassafras tree. It grows largely in dry woods and thickets, and very often fails to attain more than the height of a large shrub, never reaching more than thirty feet. The plant is easily identified by the interesting greenish branchlets, and the fan-shaped leaves which, in their usual bisected form appear like a mitten with two thumbs. Almost every part of the plant is useful in one way or another, but at the moment it is the dried leaves which are the source of the commercial sassafras. These powdered leaves provide the basic ingredient, in the South, for the "gumbo filé" made so famous in New Orleans restaurants.

As with most plants which have a number of uses, the plant has acquired a variety of names in this country, among others being that of ague tree, saxifrax, cinnamon wood, saloop, smelling stick. The first one of these at least, indicates the fact that the tree has been valued for its medicinal qualities, and, in fact, it was the hunt for

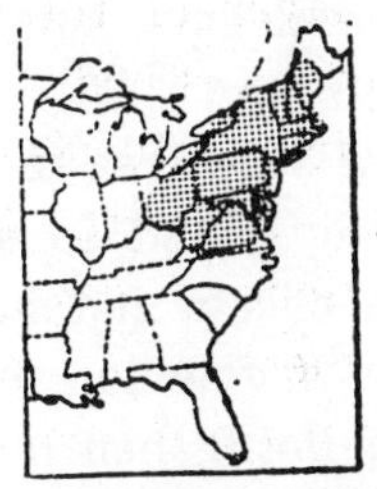

this tree which inspired early American exploration. Gosnold, in 1602, felt that he had achieved the objectives of his voyage when he found great quantities of sassafras growing on Martha's Vineyard Island, just off the southern coast of Massachusetts. Sassafras is no longer of great importance in the pharmacopoeia but there is still a moderate demand in medicine for the extracted oil, which goes under the name of *Oleum sassafras.* In any case, it is quite likely that the aromatic and stimulating properties of the juice of the roots was felt to be medicinal in the early days, and perhaps had values which were psychological if not physical.

Another use for the sassafras in colonial times was as a source for dye. Depending upon the quantity of the dye used, shades varying from pink to a warm brown are obtained.

Any reader of this, wishing to experience the flavor of sassafras can make a tea by boiling the roots in hot water, adding sugar when the brew becomes a deep red color. Unless for some tastes it is too much like toothpowder, it may be considered a palatable drink. Because there are definitely stimulating qualities to the plant, it should be remembered that any overdose of the essential oil of the plant could be narcotic in action, but this is unimportant in ordinary kitchen use where it is valuable partly for its flavor, but mostly for the thickening and mucilaginous quality which imparts a real smoothness to many culinary preparations.

Other commercial uses of the root of sassafras include the use of the oil for flavoring gumdrops, and in soap. The fragrant wood may also be put in chests and boxes, as a protection against moths. Another suggestion would be to use the wood as a material for building hen houses as the pungent oil would keep out insects.

The Department of Agriculture says that the dried bark of the root is, after some three centuries, still in reasonably constant demand for medicinal uses. In preparing for market the outer layer of the root is discarded, the rest being carefully dried. This dried root is then used for the production of sassafras by distillation, in the limited areas where it is collected. Leaves are to be carefully air-dried, then crushed and powdered.

"See thou bring not to field or stone
The fancies found in books:
Leave authors' eyes, and fetch your own.
To brave the landscape's looks."

— EMERSON, "Waldeinsamkeit"

Chapter IX
LOOK AROUND —— USEFUL SHRUBS

Amelanchier (various species)

THIS MEMBER OF THE ROSE FAMILY has a number of common names, such as Shad-bush, June Berries, Service Berries, or Indian Pear, and in this country the fruit has long been used as a source of food. The Indians used the dried berries as a basic ingredient in their *pemmican.* The rather pulpy fruits were gathered in large quantities, pounded, dried, and then mixed with corn meal, buffalo meat, or other foods to make their pemmican cakes.

The pulp of the ripe fruit in almost all of the species of Amelanchier is quite juicy and sweetish. Although not as valuable as many other plants of the wild for culinary purposes, the smaller fruits can be used in muffins, cooked in a pie as "mock cherries" or dried and used as "currants."

In the early spring, the five-petaled flowers are abundantly produced, and make this shrub one of our most welcome. Its flowering in the tidal river areas usually coincides with the "running of the shad" and hence one of its common names. The tiny apple-shaped fruits are not usually ripe until July or August; according to the species, they may vary in shape from round to pear-shape, and in color from deep red to purplish and blue-black.

If in doubt about the identification of these fruits, the ten seeds can be counted, note also being made of the smooth grayish bark of the young growth, and the coarsely toothed alternate leaves.

The Amelanchiers vary in growth from four-foot shrubs up to thirty-foot trees, and as a rule are to be found growing in open situations in poor soil on the borders of woods. In one form or another, they can be found throughout the northeastern states and ranging as far south as Louisiana.

SHAD BUSH — SERVICE BERRIES

Arctostaphylos Uva-ursi — BEARBERRY

One of the most interesting facts about this plant is that its scientific name is a repetitious one, both the generic and specific names meaning "Bear." In the sandy and probably somewhat acid soils in which it grows, it is very common in our territory, and has been admired for so many centuries that it has more common names than most plants, including such interesting ones as: kinnikinick, mealy berry, mealy plum, hog cranberry, bousserole, bear's billberry, bear's whortleberry, foxberry, rockberry, crowberry, mountain box, barren myrtle, the universe vine, brawlins.

Looking at certain of these names it will be seen that its use as a food for wild animals has long been known, while other names connote its possible use as an ornamental plant. Indeed, the tiny glossy leaves of the bearberry and the bright red fruits which follow in the fall make this an excellent ground cover plant, especially where proper growing conditions can be maintained. One plant of it may run for a great distance, making it very difficult to transplant.

Except to wild animals, the food value of the plant is quite unimportant, but the medicinal values of bearberry have been known for centuries, especially for diseases of the urinary tract. For such medicinal purposes the green leaves may be picked in the fall, thoroughly dried, and reduced to a powder.

Berberis vulgaris — BARBERRY

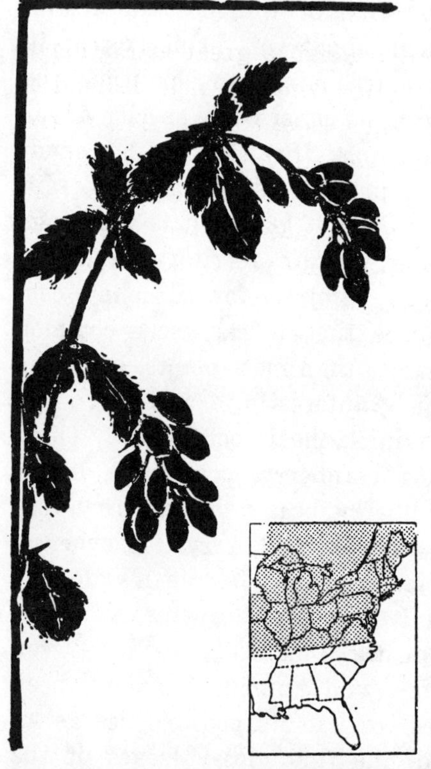

Although this plant has been legislated against for several decades — and has probably been quite well eliminated from most sections of the West where wheat growing is important — isolated plants of the barberry are to be found along almost any country road in the Northeast. To the wheat grower the barberry is a great enemy, because it is a host plant of a fungoid wheat disease; but for those who are not engaged in farming, the barberry is indeed a remarkably useful plant. The barberry shrub can easily be identified by its drooping branches and egg-shaped red fruit, and when used as a specimen lawn plant it can be very beautiful.

There are existent records from the time of the Egyptians telling of various uses of the barberry, especially of its medicinal use for liver complaints. The yellow stems of the plant produce a dye which is permanent, without the addition of mordanting materials. The slightly acid fruits which are so abundantly borne in the fall have a prime culinary use in the making of spiced preserves. The juice in jelly-making may be extracted without adding water, by cooking slightly, and if carefully strained makes a tart and distinct flavored jelly.

Fruits of the barberry are also a great attraction to birds; especially the Cedar Waxwings, which descend in great numbers on the large bushes during the fall migration. As a shrub it should certainly be planted in any bird sanctuary, and with all the other values added, the common barberry will find a top place on the list of useful and interesting plants of our area.

Ceanothus americanus — New Jersey Tea

This is a relatively unimportant plant which grows as a sprangly shrub in dry open woods and on rocky banks. It can be identified by the alternate oval leaves, very closely edged with fine teeth, with three ribs running from base to the apex of the leaf. Its greatest claim to fame is that it seems to have been used to some extent during the American Revolution as a substitute for oriental tea, to which it has but small resemblance and none of the exhilarant properties. After the Boston Tea Party many substitutes for tea were tried, and this was one of the more commonly used, especially in the southern parts of the country.

Clethra alnifolia — SWEET PEPPER BUSH

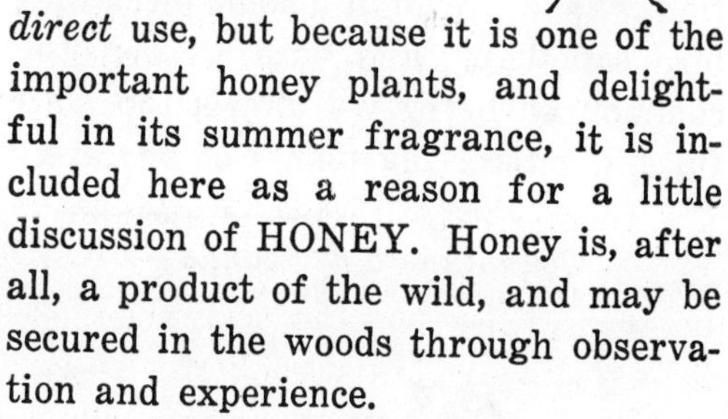

This is (intentionally) the only plant in this book for which there is no *direct* use, but because it is one of the important honey plants, and delightful in its summer fragrance, it is included here as a reason for a little discussion of HONEY. Honey is, after all, a product of the wild, and may be secured in the woods through observation and experience.

The locating of a supply of honey produced by bees "gone native" is the means of providing a major hobby to a small group of people in this country. An exponent of bee-hunting was the late George H. Edgell, Curator of the Boston Museum of Fine Arts, who has developed the locating of swarms of bees in the wild to a fine art. This has been discussed by Mr. Edgell in his booklet, *The Bee Hunter,* (Harvard, 1949) and he thus describes the joys of his hobby —

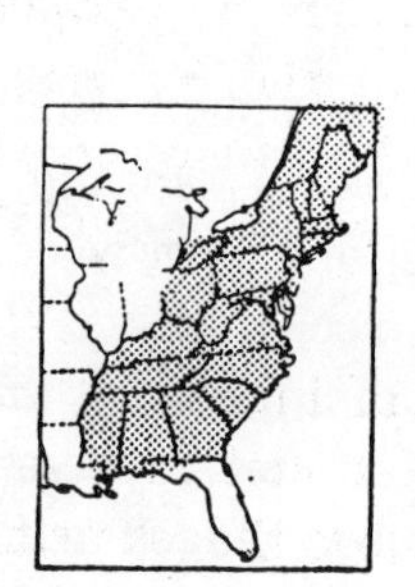

"Bee hunting is one of the most fascinating of sports, and one could go on describing different illuminating episodes for many pages. The sport combines almost everything that is desirable. It is played out of doors. It requires

> exercise both of the muscles and the brain. It is a sport of brawn and of craft. It can be played alone and is one of infinite variety, of suspense, disappointment, perseverance, and triumph. You go out into the field. Before you is a bee tree. — The reward is when, after hours or days of trial and error, your eye catches the flash of wings in the tree and once more you are able to say checkmate in one of the most difficult, complicated, and fascinating games in the world."

To go on to discuss the history of bees and of bee culture and all the properties of honey as well as all of the interesting lore written about bees is quite outside the province of this book. It is said that more books have been written on bees than about any other creature except man. We know that the Bible contains many references to the product of the hive, and that all through human history honey has been in some way connected with commerce and the finer things of life. Beeswax enters into religion as a most desirable material for making church candles, and the name "honey" is a preferred term of endearment as well as a symbol of everything that is pure and wholesome.

Medicinally, honey is important as an ingredient of cough syrups and is often prescribed as a diet in cases where no other food will be accepted by the system. Further it is said that honey has the property of killing germs and thus is always pure. It is furthermore utterly stable and long-lasting.

In addition to the sweet pepper bush as a source for honey, there are a number of other plants listed in this book which provide nectar. This includes clover, members of the mint family, and to a considerable extent, the flowers of many of the trees and shrubs.

Comptonia peregrina — Sweet Fern

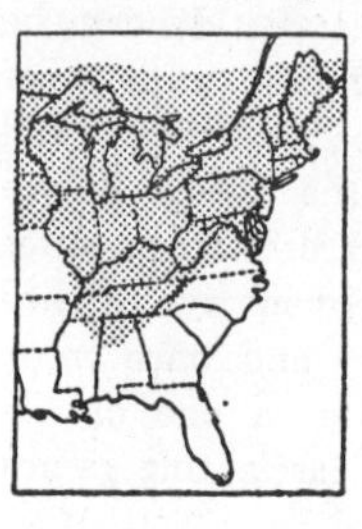

This is one of that very small group of plants which will grow on ground which seems to be utterly sterile. As found, it is a small shrub, rarely growing over three feet tall, with elongated toothed leaves; in every way a shrub quite delightful and distinct.

The aromatic leaves make a palatable tea when dried, and the young nutlets are nice to nibble on in early July.

Although not now highly rated for any medicinal properties, at one time considerable use was made of a decoction of the leaves for diseases requiring medicine of an astringent quality, considerable evidence being given of its value in cases of diarrhea. One old book indicates that during a "bloody flux which prevailed as an epidemic in Rhinebeck (New York) in 1781, and swept off the inhabitants daily, an infusion of sweet fern was employed with such success, that it was considered almost a specific. It produces perspiration without increasing the heat of the body." Another well authenticated use for the plant has been as a remedy for worms.

It seems to the author that its greatest value, however, is in its fragrant properties, which, brought into a sick room, would dispel unpleasant odors. A handful of the leaves stripped off as one walks through the sterile hillsides where they grow is stimulating and pleasant.

Ilex verticillata — WINTERBERRY

An alternate common name for this plant is Black Alder, but a name which should raise it in public esteem would be "deciduous holly," which is what the plant really is. Certainly anyone who has seen isolated berried specimens of the winterberry laden with red berries in the late fall will agree that it is much more beautiful than our own American holly, at least from the standpoint of red berries.

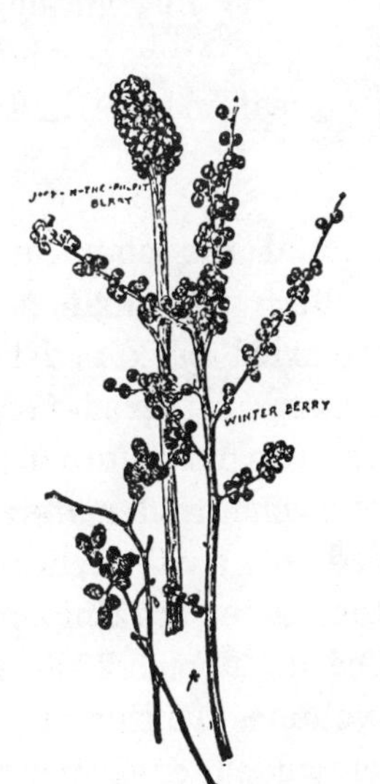

There has been some demand for the harvested berries for medicinal use, according to the U.S. Department of Agriculture, but by far the best use that can be made of this plant is for decorative purposes at Christmas. Bunches of the berries are cut and sold by boys at roadside stands in many parts of the country. Such cutting off of the upper branches to use for cheerful decoration does no harm to the shrub itself, for this pruning encourages new growth. Conservationists, therefore, should raise no great objection to the cutting of the winterberry for personal use nor, for that matter, for commercial sale.

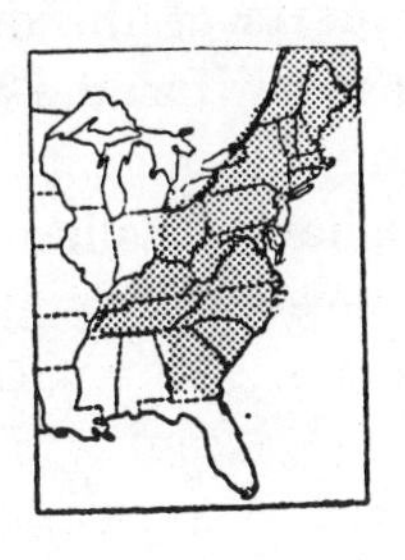

Hamamelis virginiana — WITCH HAZEL

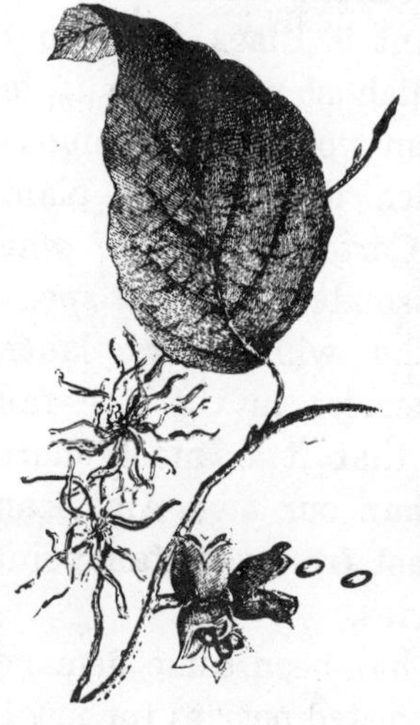

The very mention of the name of witch hazel brings to mind the possible uses of this common American shrub. It is a native of the damp woods from Maine to Florida, and is one of the few woody plants which produce flowers in the very late fall. The dull yellow flowers are somewhat similar to that of Forsythia, although not as noticeable, except where they grow in great masses in the deep woods.

The parts of the plant used commercially for medicinal purposes are the dried bark and the leaves, which are gathered fresh and then dried. In home practice, a quantity of such dried material can be added to two parts of water, and then after soaking, the liquid distilled to about 25% of the original volume. To this distillation is then added one part of alcohol to five parts of the distillation, the end result being a very stimulating and astringent rub, which has peculiarly healing and pain relieving powers.

Gaylussacia baccata — HUCKLEBERRY

Gaylussacia frondosa — DANGLEBERRY

DANGLEBERRY

With the true blueberry so abundant throughout the United States, its cousin the huckleberry is not much appreciated, delicious as it may be. It grows in much the same circumstances as the blueberry (which see), and variably may be found growing in very dry and rocky places or in extremely boggy ones, this like other members of the Heath family requiring above all else an acid soil for complete happiness.

In any case, the true huckleberry does make very delicious pie, the only reason for its apparent lack of popularity being the ten small, stonehard seeds which are found in each fruit. Actually the huckleberry has a spicy sweet flavor, which to many tastes makes it superior to the blueberries are so often found growing in similar locations, either one can be picked to advantage, and where there are no blueberries, the huckleberries certainly should be gathered as a flavorful substitute.

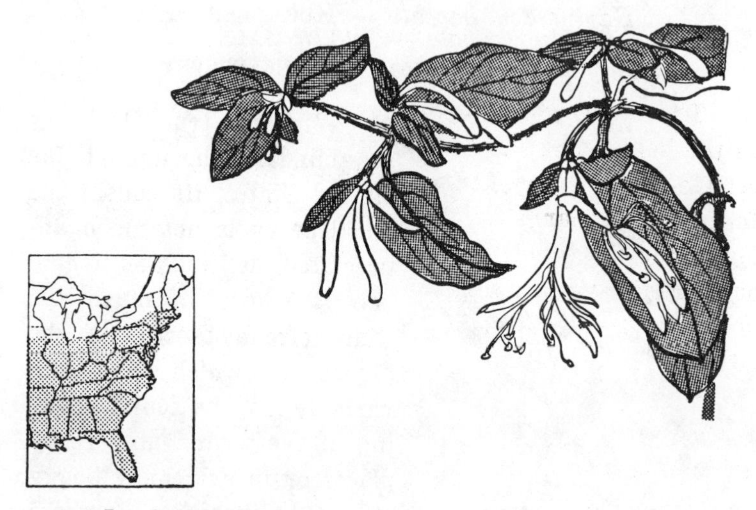

Lonicera japonica — JAPANESE HONEYSUCKLE

So wide-spread has the use of this vine become in the United States that it is almost a wild plant, and the more especially as it establishes itself so firmly, once planted. It is included in our list because the long woody stems of the plant can be dried and used for making simple baskets and ropes.

It is not hard to identify the Japanese honeysuckle, because of the almost evergreen leaves and the white, changing to deep cream, flowers which are produced freely throughout most of the summer. Indeed there is nothing quite so delightful as the wind-wafted fragrance of the honeysuckle on a humid summer evening.

The honeysuckle (as mentioned in another chapter) is also valuable in landscaping for holding banks and for providing shade on a porch. It is an extremely versatile and hardy plant which has come to us from the Orient, and which through one means or another has now become widely naturalized.

Myrica gale — SWEET GALE

The sweet gale (also known as the bog-myrtle) is very closely related to bayberry, and will be found in similar situations as a low shrub (4 feet) growing in damp acid soils throughout most of the colder sections of our territory. The leaves are like those of the Sweet Fern (*Comptonia peregrina*) except that they are not lobed. The fruit produced by the plant is in the form of a nutlet, which has been considerably used in some countries, (France for instance) as an aromatic spice, for it has a biting taste somewhat like sage.

In other places the leaves also have been cured for making a delicate tea. In Maine, the plant so used goes under the name of meadow-fern, as a cure for colds; but in view of recent researches on cold cures, its value here must be doubtful. One authority suggests that the resinous fragrant leaves are (like red cedar) useful to keep moths out of dresser drawers. Another writer reports that the Indians used the young buds of this plant as a source for dye for coloring the ornamental porcupine quills used in their costumes. Consideration of such a plant as this well shows how imagination and necessity will find uses for very inconspicuous plants.

Myrica pennsylvanica — Bayberry

Bayberry is one of the most familiar plants of the sterile coastal plains. It is noticeable especially as a shrub with gray and stiff branches, clothed with erect growing leaves, two to three inches long. The terminal fruiting branches are covered early in the fall and winter with bunches of round, roughish and dull, whitish-gray berries. Identification can be further made by the strongly aromatic smell of the leaves when crushed.

The principal use of the bayberry, as gift-shop patrons know, is for the production of candles. Although an early American herbal glibly says "there is perhaps no form of disease in which the bayberry, if properly administered, will not prove beneficial," any medicinal use of the bayberry today is rare.

In the absence of the leaves of the true tropical bay tree, the leaves of this plant may be used in cooking, as a substitute for "Bay Leaves" but this is usually unnecessary. A much better use of the bayberry is as a dwarf landscape shrub. So used it will quite likely attract the myrtle warblers, for whom the berries are a special treat. Not only does the myrtle warbler feed on these berries, but ornithologists say that some ninety other species of birds will use the bayberry as food.

Other uses aside, it is the fragrant wax that makes the bayberry such a desirably useful plant and instructions for extracting and using the wax in candle-making is given in the chapter on plant crafts. In earlier days, bayberry wax was considered of more than average merit for smoothing flatirons.

Ribes triste — WILD CURRANT

Plants of our native currants are seldom found in the wild. In the southerly regions may be found *R. glandulosum*, which is less palatable than *R. triste*. Almost everyone knows the currant, species of which can be easily identified by the typical serrated and generally maple-leaf-shaped foliage. The fruit is a one-celled berry filled with pulp, crowned with the remains of the old flower.

It is said that our name of currant was given to garden forms of the Red Currant because they resembled in size the South-European currants which grew near Corinth, which name had become corrupted to currant. The fruits of the wild species of currant are small and quite sour, but when found, are useful as a piquant jelly. Improved forms of this native wild currant have found their way into cultivation, all fine for the small garden.

On occasion one will also find in the country isolated plants of our native Black Currants (*R. Americanum*) which are also useful for making into jam, though not as fine as the European species. Laws in most states prohibit the growing of currants, especially the black currants, as all members of the family are hosts for an intermediate stage of the White Pine Blister, which is so disastrous to that finest of our northern evergreens.

Prunus maritima — BEACH PLUM

Because of the importance of the beach plum to the coastal areas of the northeastern United States, considerable interest and research has gone into its values for commercial, garden, and landscape use. To the writer, one of the great values of the beach plum is its beauty as an early spring flowering shrub; but because of its general unhappiness when transplanted and its preference for specialized conditions, it is not to be recommended as a landscape plant for any but sections where it is found growing naturally.

The beach plum is an extremely variable plant through natural hybridization, some shrubs attaining a height of ten or twelve feet, while other plants seem to remain almost prostrate. Again, the colors of the fruit range from purple through red and down to almost pure lemon-yellow when ripe, whereas in size the beach plum varies between a small damson plum and a small crab apple, the latter-sized fruits being of small value as they generally have more pit than pulp.

The beach plum should be harvested when the fruits are just showing their full color, and may be made into jelly or into a very tart jam. Whatever way it is made, it is a wonderful accessory to a meat course, and always a pleasant winter reminder of happy summer and fall days at the beach.

RUM-CHERRY BEACH PLUM

The Rum-Cherry
is discussed
on page 126

Large illustrations
(as above)
by Miss Frances McGaw

Rhus glabra — Scarlet Sumac

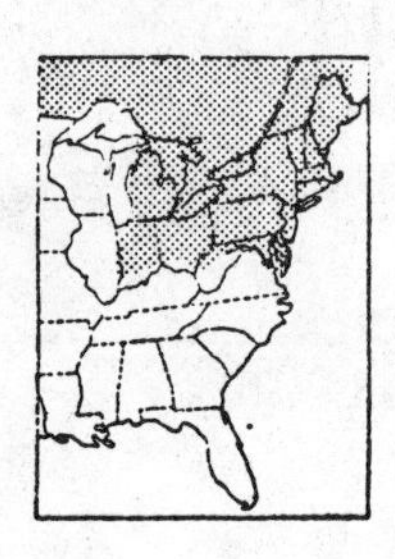

This large shrub or small tree is always an interesting feature in the landscape either in winter or summer. In winter, the skeleton-like outline of the shrub stands out strongly with its dark reddish bark, whereas in the summertime, the almost fern-like and yet tropical appearing compound leaves provide a vigorous note in any landscape.

It is in late summer, especially, that the pyramidal bunches of berry-like fruit strike an insistent note in an otherwise green woodland setting. The sumac rarely grows in dense woods, more likely being found on the edges of plantations, out along the roadside, or alone in clumps in dry open soils.

It should be noted in considering the sumac that there is one species of shrubby *Rhus* that falls into the category of poisonous plants. A first cousin of great notoriety is poison ivy. (See Chapter 13). There is, however, small danger of confusing the poison sumacs with the scarlet sumac, because of its fruit heads, and always distinctive appearance.

The element of the sumac in which we are interested is in the berries which, because of their malic acid content, are the source of a pleasant and pretty drink. The Indians gave to the early Americans the knowledge of this drink, prepared by bruising the fruit in water, straining through a cloth to remove the fine hairs, and adding sugar to taste. A drink so made resembles pink lemonade and is refreshingly palatable. So much did the Indians treasure it that they dried the berries to ensure a supply of the acid drink in the wintertime.

SASSAFRAS SUMAC

Perhaps the greatest practical value of this and other forms of the sumach is as a source for commercial tannin for tanning and dyeing. Persons interested in gathering the leaves of sumach as a little sideline should send to the U.S. Department of Agriculture for "Production Research Report No. 8 — 1957" (10c).

In some sections of the country, the name of the plant is pronounced "sho-mack," elsewhere "sue-mack;" with "sumach" a variant spelling.

Rubus occidentalis — Black Raspberry

Of all the fruits which occur naturally in the wild in this country, perhaps the best, in flavor at least, is the black raspberry. Anyone who has eaten black raspberry jelly knows that it has a flavor quite unlike anything else, and although the fruits are full of seeds, they offer a high reward to the palate. Wherever the black raspberry occurs, it produces fruit of good quality in the wild, even often superior to the cultivated forms.

The fruit of the black raspberry can be used in any way at all; fresh, cooked, preserved as jelly, made into a drink, and as a flavor for ice cream. The leaves when dried make a desirable tea, while the young tender shoots of the plants when peeled are good for a tasty nibble. Never neglect looking for black raspberry bushes in likely locations.

Within the range of the northeastern United States there are in addition to the blackberry and raspberry other forms of this general family of *Rubus,* worthy of mention being the purple-flowering raspberry. The cloudberry or baked-appleberry, principally found in Maine, is also interesting, although not as flavorful as other species.

But about any of these berries, as one writer well said,

> "There are two sequels to a successful afternoon's blackberrying which are delicious, unique, enchanting. First, the smell of a basketful of the fruit, especially if it is sun-warmed. There is nothing in all nature like it. And later, there is the jelly, and both the jelly and the smell of a basketful of berries is ambrosial and flawless."

Rubus flagellaris — THE BLACKBERRY

Speaking botanically, it is perhaps somewhat presumptuous to pick out one species of the blackberry for discussion, because there are so many variant forms of this useful wild fruit which are available for culinary purposes. Modern botanists tell us that there are several hundred species in the United States, many of which are highly variable; and it is from these that there have been created through hybridization such valuable garden plants as the dewberries, the loganberries, and various hybrid blackberries, each adapted for growth in various climatic or soil conditions. Some of them are shrub-like, while others are creeping in habit.

For real flavor and sweetness, this writer feels that most wild blackberries are still superior. The fruits are just as easy to gather and a lot more fun, and one could perhaps truthfully say that this is our most valuable wild fruit. Generally speaking, wild blackberries like to grow where there is sunshine, and they will be found in open fields, on the edges of copses, or along the roadside.

In picking blackberries, one should take time to note the variability of sweetness and flavor in the various colonies in which they grow, and select only those fruits which are the largest and best flavored, since some of the small natural hybrid forms have fruits too seedy to take home. Blackberries are at their best when simply put down as jam with equal parts of sugar, as cooking somehow seems to augment the flavor.

But it is not food value alone for which the blackberries are noted, for the growing shoots are a source of dye, while one of the commonest forms of medicine known to our ancestors was blackberry brandy. If you are one who likes to make the fullest use of native plants, it will be fun to prepare a supply of the useful medicine which the blackberry offers.

The entire plant contains an active principle which for centuries has been known to be about the most certain cure for diarrhea. The roots of the blackberry (and the leaves as well) are dug and dried, and then stored until wanted. To prepare the medicine for use, a teaspoon of this dried, crushed root is stirred into a cup of boiling water. Drink one or two cups when cooled, until the condition is remedied. Probably one would find that the eating of a large quantity of the fresh fruit in summer would have about the same effect, because every part of the plant contains the active principle which is useful.

Turning now to a much more questionable medicinal use, one herbal makes the statement that the young shoots eaten as a salad will fasten loose teeth, and that a decoction is useful against whooping cough, while the more ancient herbalist, Gerard, says that "a bramble leaf heals the eyes that hang out." The Greeks, according to one authority, used the blackberry as a remedy for gout.

What use, other than medicinal, one might make of blackberry cordial or brandy will depend on the attitude of the reader towards the social aspects of the use of alcohol.

The species BLACKBERRY *is highly variable, but to the amateur, most forms look like the illustration opposite.*

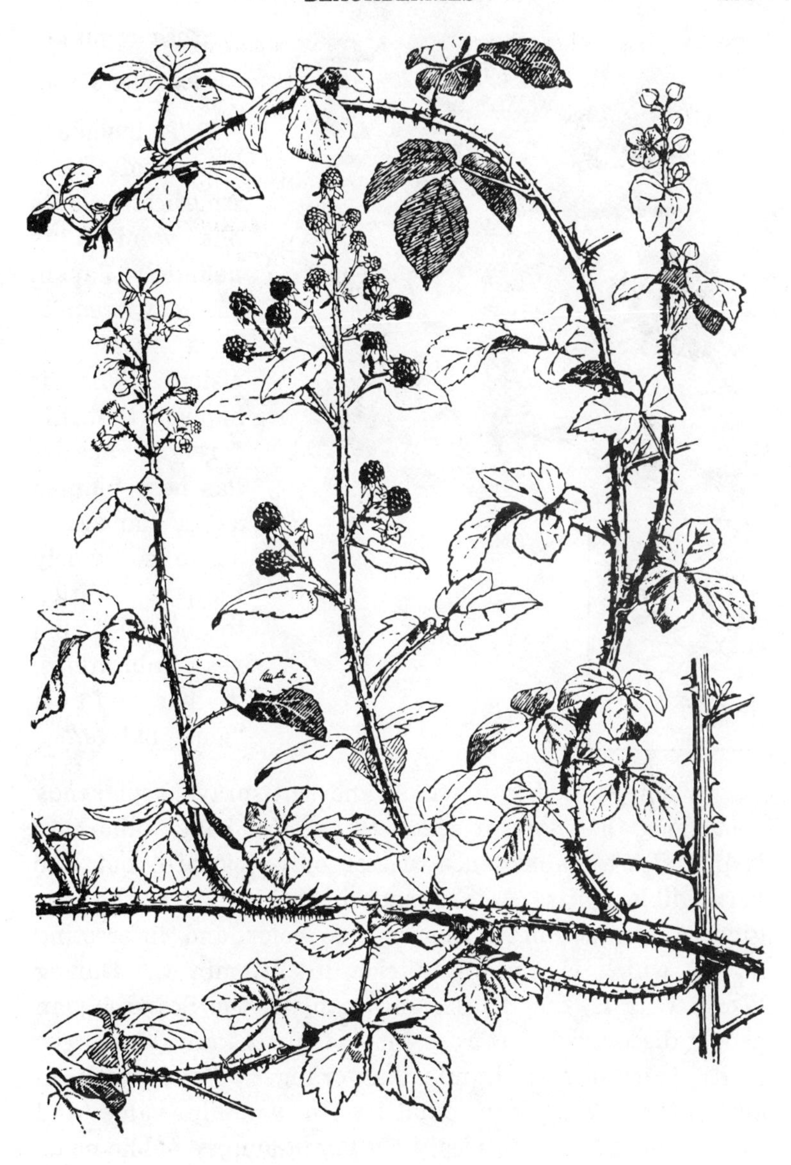

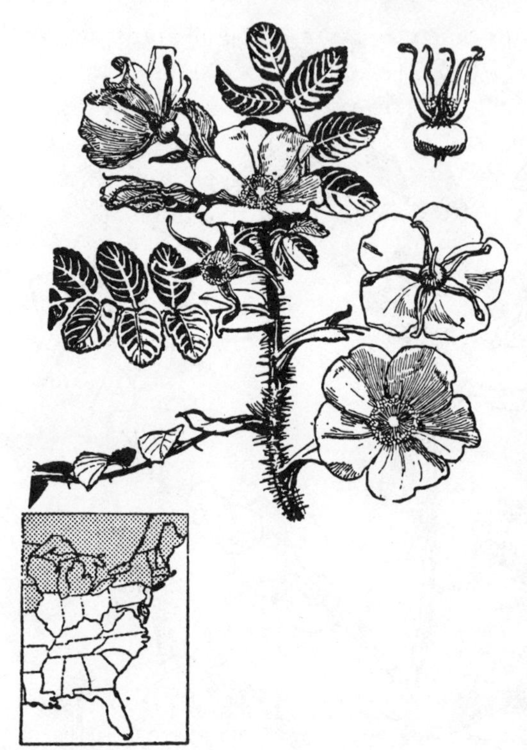

Rosa rugosa

JAPANESE ROSE

To include in a book of useful American plants one which is definitely Japanese may seem to be a strange decision. But it happens that this foreign species has been planted as a landscape shrub so widely that, in addition to being found near habitations, it has often "gone native."

When planted in a windy and salt-spray-swept sandy spot along the shore, it seems to be most happy, and here it provides an abundance of not only beautiful but also very edible fruits. A jam made by boiling down the hips gives a product of a rich orange color and interesting flavor, which is amazingly rich in Vitamin C. During World War II, the British and many of the Scandinavian people discovered that rose hips of almost any kind were much richer in this vitamin than oranges. For this reason authorities gave wide publicity to rose-hip value and throughout those countries the vitamin-hungry public made much use of the rose fruits as a substitute for the then utterly unattainable citrus fruit.

The rather bland flavor of rose-hip jam needs a little spice to make it entirely desirable. If one likes it may be flavored with rose water, or a leaf of rose geranium placed in the bottom of each jelly jar will provide a substitute for rose essence. It may be noted that the seeds or hips of other than Japanese roses may be used in this same manner.

The value of *R. rugosa* as a landscape shrub, especially in sea-shore planting, is well known.

Juniperus communis — Ground Juniper

This shrubby, warm, gray, evergreen is a native of the sunny barren hillside throughout much of our territory. It rarely grows over four feet in height, more generally three. It is characteristically found with its close relative, the Red Cedar, the two together forming massive landscape groupings that could not be improved on by the best landscape architect in the country.

Considering its prickly leaves, about the only good thing that can be said for this juniper is that the berries are a common and essential ingredient of gin, to which potent drink the juniper imparts a characteristic flavor. Some authorities suggest that Juniper Berries might be more often used as a seasoning in cooking, as the flavor is somewhat similar to the tropical Bayleaf. In some parts of the world the berries have been roasted and used as a substitute for coffee. One authority on herbal medicine recommends that an infusion of the dried ripe fruit be used as a diuretic and stimulant.

Sambucus canadensis — ELDERBERRY

So widespread are various forms of the elderberry and so useful is one or another part of the plant that it is hard to select the most important uses to discuss. One could enumerate them as follows:

> *Culinary*: The berries as a superior pie-fruit; the flowers and berries both as a source for wine.
> *Medicinal*: Some listed values.
> *Dye*: The berries as a source of deep-red color.
> *Crafts*: The hollow stems are useful in many ways.
> *Landscape*: The elderberry is highly useful as a flowering shrub.

The beautiful white heads of the flowers of the elderberry are bright in the landscape in early summer, these being followed by the nodding clusters of fruit, varying in color from red to black. The plant is found on roadsides, hedgerows, but most happily, somewhere near water, where the elder makes a rank growth and produces superior fruit.

The best use for elderberries is for pie and next as a source for a delicious wine. Whether the pale yellow wine which comes from the flower heads called "elder blow wine" or that from the fruit is the better is for the epicure to decide, but they are both good when properly made. Instructions for these are given in Chapter II, but it may be said here that the principal thing to watch in the making of elderberry wine is that every process should be conducted with great care, to avoid spoiling the wine in the early stages through the introduction of vinegar ferments.

THE ELDERBERRY

So old, also, is the tradition of the medicinal usefulness of the elder that recipes for such use crop up perennially in the garden magazines. One of these from England recently detailed the making of an ointment from the flowers to heal the sores on animals, while in that reliable farm journal, the *Rural New Yorker*, a similar ointment is recommended to "beautify the complexion." The latter recipe goes something like this:

A quart of lard is melted, and into it is put as many elder flowers as the lard will cover, all being cooked slowly for an hour, then strained and poured into small jars. For bruises and sprains, a more potent ointment is recommended to be made from the young elderberry leaves, while a syrup made from the berries, was suggested for coughs and colds.

As the writer of the above formula said, "Whether or not present day medicine agrees with the remedies above as being really effective, they were the homely recipes used in days gone by. We do not set them down here with any idea of cures, nor do we ask you to follow them. They do show, however, what our forebears turned to in their own times."

So as a source for a boyhood whistle or the wherewithal to wet one's whistle in later life, there's always the elderberry bush.

Viburnum cassinoides — WITHEROD

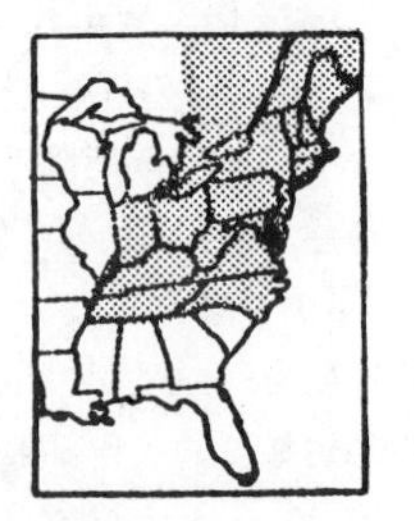

Somewhere growing on the edges of streams or swamps one will find the plants of this species of the viburnum. It grows as a large shrub with terminal clusters of flat-topped white flowers, which are followed by blackish purple plum-like fruits about the size of a raisin. These fruits may be eaten as a nibble, while the shrub itself may find some use for transplanting into the more distant parts of the home landscape. It is not one of our most useful wild plants, but one interesting to know and observe.

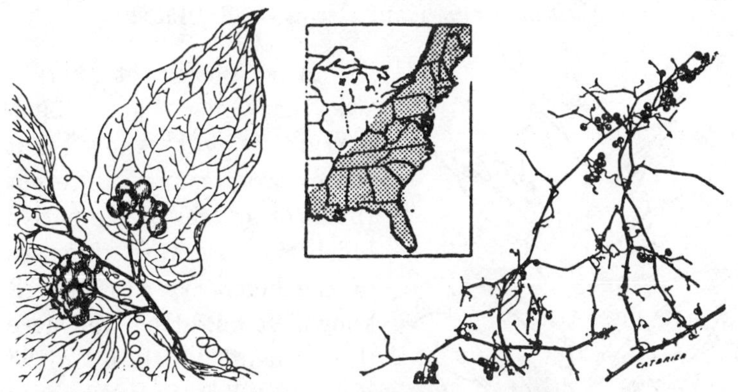

Smilax rotundifolia — BRIER

Whatever prefixed names there are to *Brier* (and there are many), the fact remains that this is one of the great deterrents to hiking in the woods. The plant grows as a coarse branching vine, rambling around through shrubs and trees, and where it grows thickly, it defies intrusion by means of its extremely thick thorns. The plant is easily identified by these two characteristics: the oval to roundish leaves, and the small black berries which appear in the fall. There is another species besides *Smilax rotundifolia,* (*S. gracca*) and between the two they extend widely in our territory.

To anyone who has cursed the brier in going through the woods, it is good to know that there is some possible value to the plant.

The berries of the brier are useful in the manufacture of dyes in shades of blue, violet and purple, while the tender young shoots when collected in quantity are rated as a delicious vegetable, to be prepared much as one would asparagus. Experiments have shown that the roots contain a quality similar to that of the pectin in apples, and that an interesting jelly may be made by boiling the roots to extract the pectin and the flavor.

Vaccinium corymbosum, et al — BLUEBERRY

Any observant person who has ever "gone blueberrying" will know that there are many different forms or species of the native blueberry. Because the uses of the blueberry are so well known, it might be well here to comment a little upon the variations of these berries which one will discover. Certain it is that the joy of berrying, plus the reward of the usually finer flavor of the wild berries over the cultivated forms, should make blueberrying one of the most rewarding of outdoor pleasures. A discriminating person will make a point of discovering the best and most desirable form of berry to gather, and thus will soon learn that even within a particular species, there is a wide variance in the size and quality of berries, with neighboring bushes requiring close selection.

In discussing blueberries one will hear people speaking of *Whortleberries, Huckleberries,* and *Bilberries* and will often wonder in what way these differ from Blueberries, because when discovered they all seem to be similar in appearance. Again one also speaks of "high-bush" and "low-bush" berries, but such a grouping also fails to distinguish properly the significant facts about the families and the variation. Without getting too deep

BLUEBERRY **BLACKBERRY**

into scientific terminology, there now follows an attempt to define the various forms:

Whortleberry: This seems to be a name now but little used and one which was given, quite largely without distinction, to the whole family of these berries, by the early English settlers, who saw in our native berries a similarity to the whortleberries (a member of the same heath family) which they knew at home in Europe or Britain. Thus "whortleberry" is more properly applied to European species of vaccinium.

Blueberry: This is a name usually given to the fruits, light blue to black in color and with a grayish "bloom." The fruit grows in clusters on the ends of branches. These slightly variable berries are know botanically as *Vaccinium corymbosum.* It is from a careful breeding with selected forms of this species that have come our many commercial varieties of blueberries. Within this genus and with a number of other specific names are found blueberries growing variously in height from three inches to ten feet, with fruit ripening from June to September.

Bilberry: This is technically just another species of Vaccinium, but one which differs from the commonly known blueberry in that it produces fruit from the centers of the leaf axils rather than in terminal clusters. It is not nearly as desirable or as perfect as the true blueberry.

Huckleberry: The huckleberry can easily be distinguished from the blueberry by the generally large and hard seeds found in the fruits. The huckleberry varies in height from three to ten feet while the fruits are black rather than blue. Some species are of exceptionally good flavor as the eaters of huckleberry pie will know, but the hard seeds make them objectionable to many. Botanically this genus is known as *Gaylussacia.* (which see).

Dangleberry: This is a taller growing species of *Gaylussacia,* in which the fruits are hung from the leaf axils on little stems, so that they "dangle." It is also known as Tangleberry.

Whatever the kind of berry, we do know for certain that all of them were favorites of the Indians. They dried vast quantities of the berries (which in this way became immune to decay), reduced this dried fruit to a powder, which then became a form of flavoring for stews. The Indian name for blueberries was "attitash." In our time, whether they be blueberries, attitash, or whortleberries, we know that, however called, there is nothing quite so popular in the restaurant or at home as good fresh blueberry pie.

Vaccinium oxycoccus — CRANBERRY

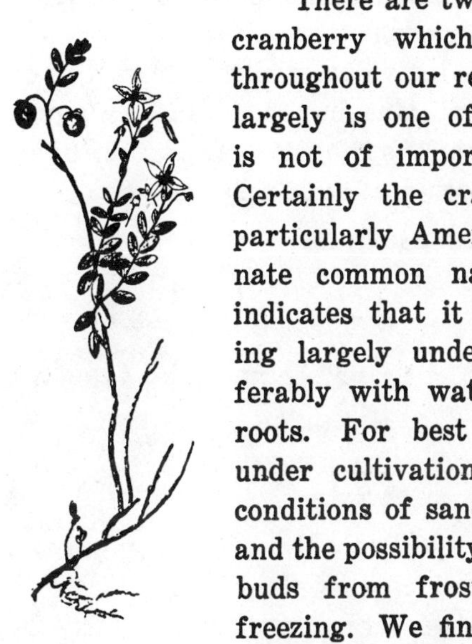

There are two distinct species of the cranberry which are found variously throughout our region, but the difference largely is one of size of the fruit, and is not of importance to the collector. Certainly the cranberry is one of our particularly American fruits. The alternate common name of bog cranberry indicates that it is usually found growing largely under acid conditions, preferably with water within reach of the roots. For best results, the cranberry under cultivation needs very particular conditions of sandy soil, plenty of water, and the possibility of protecting the young buds from frost and the fruit from freezing. We find that cranberry growing is a big and difficult business.

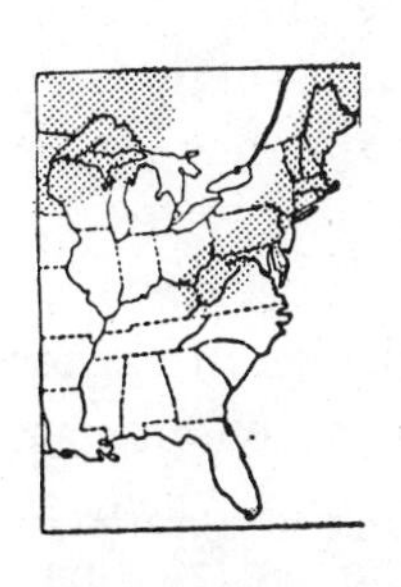

The isolated bogs of *wild* cranberries are not too easy to discover, but there are many spots where they may be found all the way from Cape Cod down the coast to New Jersey, and up in Wisconsin. The time to look for the berries is just before heavy frost, when they would be bright red in color and more easily distinguished. The writer discovered a fine growth of them one fall, within twenty feet of a very heavily traveled highway, in the back waters of a famed New England river. But they had to be hunted for, because the cranberry as a plant never flaunts itself before the eye of the public. This is so because it is a creeping plant, not over three inches tall, with very tiny glossy leaves. While the cranberry will not do well in dense shade, it does not seem to mind the protection of grasses and small over-growth.

The cranberry as an individual plant, as shown in our drawing, is utterly insignificant, but the sight of a purple-hued bog in winter or early spring is a feast for the eye of any artist and is worth traveling miles to see.

Viburnum trilobum — HIGHBUSH CRANBERRY

Probably the only way in which this plant is similar to the true cranberry is the fact that the berries are round and red and of about the same size. There are two forms of the high bush cranberry found in this general section, the one being escapes of the European variety, pictured above.

The other is the American high-bush cranberry which is often found growing on well drained hillsides in rather barren soil. It is similar in appearance to the one shown and is notable for its resistance to the aphids which so plague introduced species. The fruit of the native form has large seeds, suggesting that it is best used as a strained jelly.

In the home landscape where a large growing shrub is needed, the native high bush cranberry can often be used with great effect, both because of its flowers and the bright berries which stand well above its bold foliage.

Vitis Labrusca — GRAPE

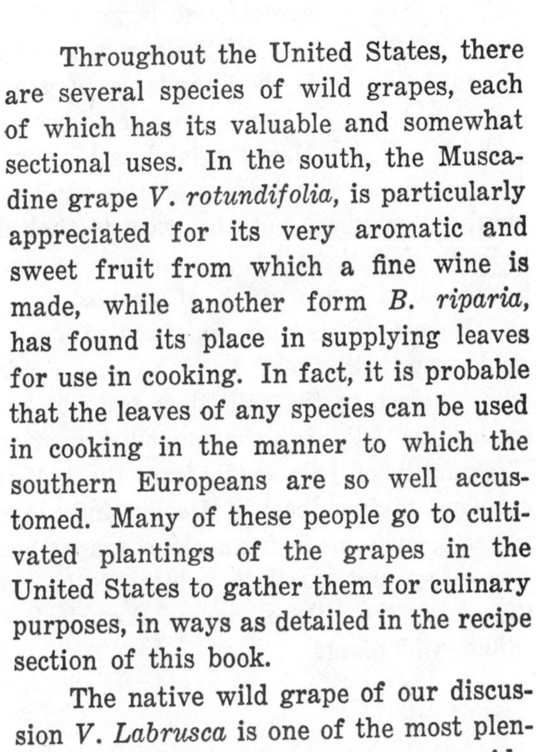

Throughout the United States, there are several species of wild grapes, each of which has its valuable and somewhat sectional uses. In the south, the Muscadine grape *V. rotundifolia,* is particularly appreciated for its very aromatic and sweet fruit from which a fine wine is made, while another form *B. riparia,* has found its place in supplying leaves for use in cooking. In fact, it is probable that the leaves of any species can be used in cooking in the manner to which the southern Europeans are so well accustomed. Many of these people go to cultivated plantings of the grapes in the United States to gather them for culinary purposes, in ways as detailed in the recipe section of this book.

The native wild grape of our discussion *V. Labrusca* is one of the most plenteous bounties of the American wayside, the vines running here and there over other vegetation along many miles of roadside. It is from this fine American species of grape that have come such excellent selected forms as the Concord and Niagara grape, as well as many newer and less well-known varieties. For this fine native grape we should be very grateful, for the European forms (now so generally grown in California), have never been happy in the East; nor in-

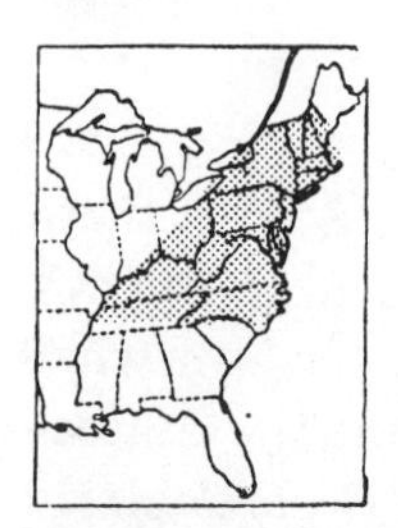

deed, even though beautiful to look upon, are they so large or flavorful.

Historians generally agree that it is the abundance of grapes growing wild which gave rise to the early name of the New England as Vineland. Not too much is known of the exploration of America by the Norsemen, but the records that do exist seem to indicate that these hardy explorers were much impressed by the luxurious growth of grapes.

However this may be, there is a great abundance of wild grapes on every hand on our roadsides and, when ready to use in the fall, they provide an easily gathered source for jelly for the embellishment of winter meals. The quality of jelly made from the half-ripe grapes is of excellent texture, beautiful color and piquant flavor, while spiced grapes made from riper grapes are quite superior. The many and excellent values of the grape will qualify it to bear the title of one of America's most useful and edible wild plants.

Chapter X

LOOK DOWN —— HERBACEOUS PLANTS

"And God said: 'Behold I have given you every herb bearing seed, which is upon the face of all the earth, . . . to you it shall be for meat.'" GENESIS I, 20

Aralia nudicalis —
WILD SARSAPARILLA

There are two species of this aromatic plant which are useful. The first is called the American Spikenard (*Aralia racemosa*), which can be identified by the cylindrical and often branching umbels of purplish berries. Some tribes of Indians apparently used this spicy-aromatic plant as a flavoring for sundry dishes and there are some reports of the berries having been used to make jelly.

More common is the plant illustrated (*Aralia nudicalis*) or Wild Sarsaparilla, the roots of which are an ingredient of root beer. It is said that the Indians used the root of this plant as a sustenance in times of famine. It is also used at times as a constituent of cough syrup with white pine and other ingredients, and as a substitute for the true sarsaparilla.

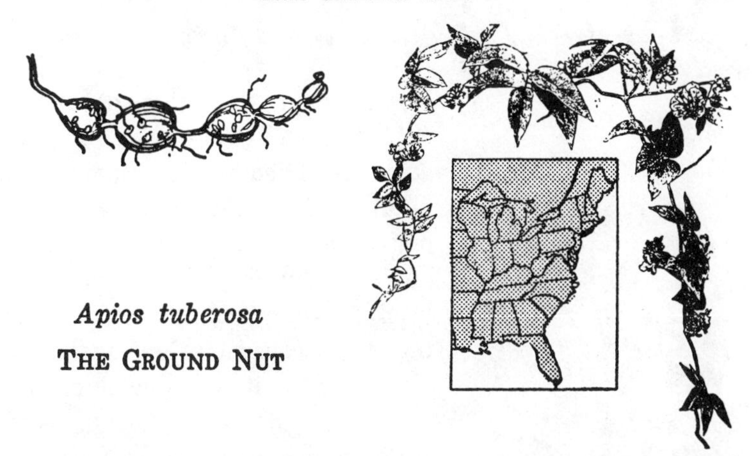

Apios tuberosa

THE GROUND NUT

Everywhere throughout our territory, one can find this quite unusual wild plant growing in the woods. In the summer time it can easily be identified by its twining vines, the soft green stems (which become dry and whitish in winter), the alternate leaves, and when in flower, by the chocolate brown fragrant flowers. These flowers are followed by pods resembling beans, which, like the tubers, are edible. The plant naturally does best in the rich soil of old woods, chiefly near a supply of water.

The most edible part is the underground tubers, which grow somewhat like potatoes in strings of smallish tubers just beneath the surface of the soil. This ground nut when dug in the fall is perfectly safe to eat in the raw state, but it is made much more palatable by roasting, after first parboiling with salt; or better yet, the small unpeeled tubers may be cut into small pieces and fried raw like potatoes.

The early explorers of the United States mentioned the ground-nut, Gosnold in 1602 reporting it as one of the valuable plants on Martha's Vineyard. Some have said that these then abundant tubers saved the lives of the Pilgrims during their first years in New England.

Arctium minus — COMMON BURDOCK

Anyone who has had contact with the annoying seeds of burdock or tried to eliminate the plant from a garden area would doubt that this plant could have any possible uses. We have all been unwilling agents in scattering the seeds of this unsightly plant, and without question, because of the annoying hooked scales which hold the seeds, it is one of the most widely-known of all wild plants.

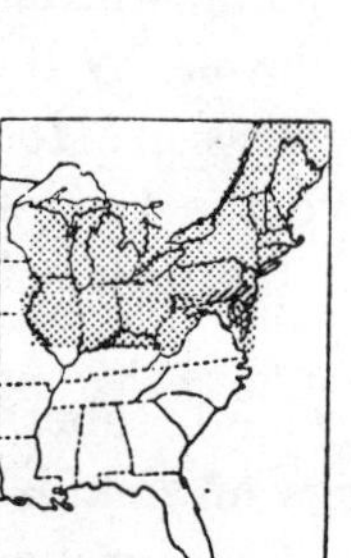

There is, however, quite a body of literature testifying to the values of burdock, both for eating and as medicine. The part of the plant used for food is the young flowering stems, as well as the pith of the root.

If these parts are carefully peeled and boiled in two waters, a flavorful and valuable "green" is the result. It is said that the Iroquois Indians dried the roots of the plant for use in the winter.

However that may be, the more we develop a taste for the cooked burdock, the sooner the plant will be exterminated from many back yards in this and other countries.

Allium canadense — WILD ONION

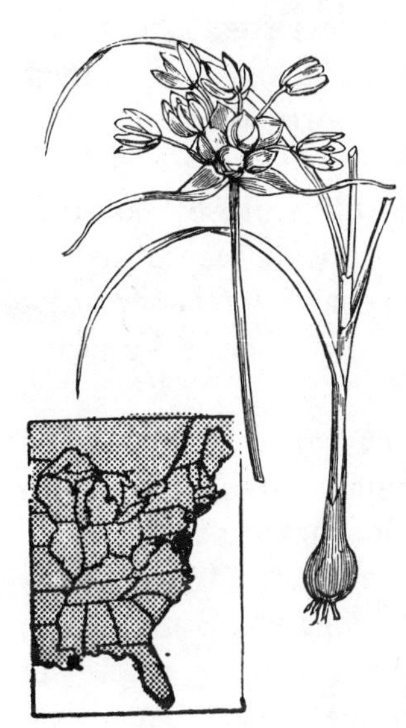

One often finds the wild onion growing in rich meadows or thickets, recognition being made easy by the characteristic onion odor and the slender quill-like leaves. The bulbs are the useful part of the plant, and although they are very small and the supply often limited, they are very sweet and palatable. One may, indeed, use the entire plant for cooking purposes, preparing from the cooked onions a delicious onion soup. The bulbs for cooking are found in late autumn or early spring.

Atriplex (*species*) — ORACH

Without close observation, one would assume that this was a form of the common Lamb's Quarter or pigweed, and botanically speaking, such is nearly true. Differing from *Chenopodium album* (which see) which grows in unpromising and dry areas, the orach is found at its best growing on seashores from Maine to Virginia. The succulent leaves of plants growing thus along the shore are juicy and somewhat impregnated with salt.

Asclepias syriaca — Milkweed or Silkweed

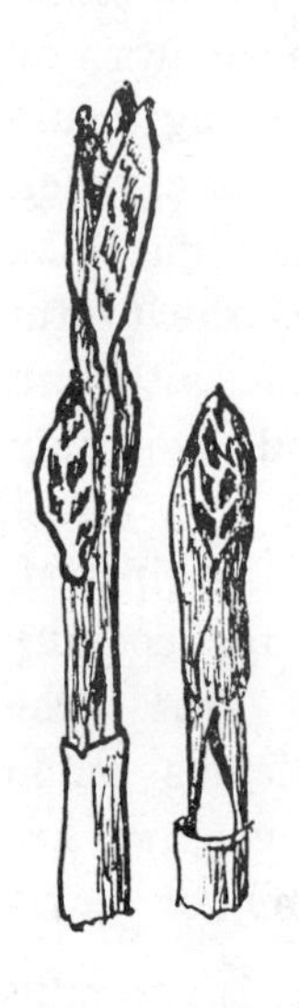

Much has been written about the useful qualities of milkweed as a food plant and especially are all authorities agreed that the young shoots of this one species are a good substitute for asparagus. Tender shoots should be cooked in several waters (to remove the bitter principle) and served with butter and seasoning.

Just because the name of the family is milkweed, identification of the plant should not be made on the basis of its milky exudation, because other plant families which are inedible, and indeed even poisonous, exhibit a milky sap when bruised. But combine milky sap with rather stout stems and opposite-ovate leaves, and identification becomes fairly easy. This commonest form of the milkweed is usually found in dry open places on roadsides or along fence rows, and should be looked for in the late spring.

The empty dried seedpods of the milkweed are extremely decorative and are often seen in florist shops painted or gilded, while the silk-like seeds have a value in craftwork. The milkweed is altogether a desirable and beautiful plant at any season and should be more generally known and studied.

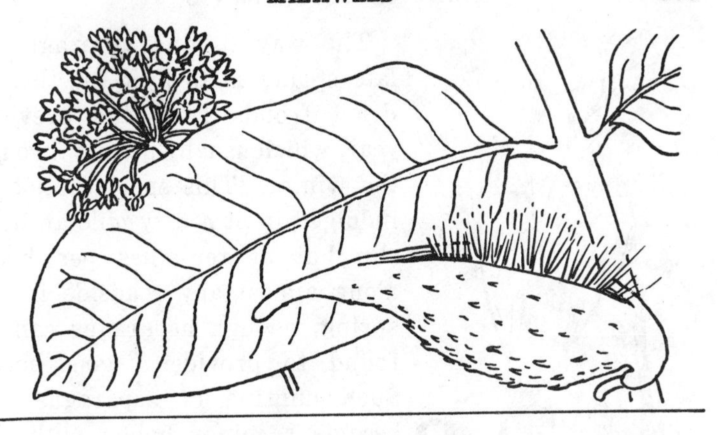

Asparagus officinalis — ASPARAGUS

This is one of our immigrant plants which has become quite as happy here as have many of the people who brought it to this country. Asparagus is a native of eastern Asia, now naturalized widely in many parts of the world. One finds it commonly growing throughout this area, along the roadsides or hedgerows, seeds having been carried there from gardens by the birds. The author has gathered asparagus from the hedgerows since boyhood, and is ready to state that the wild is quite as good as the cultivated, it being, of course, the same plant. Very large stalks may be obtained from these roadside plantings.

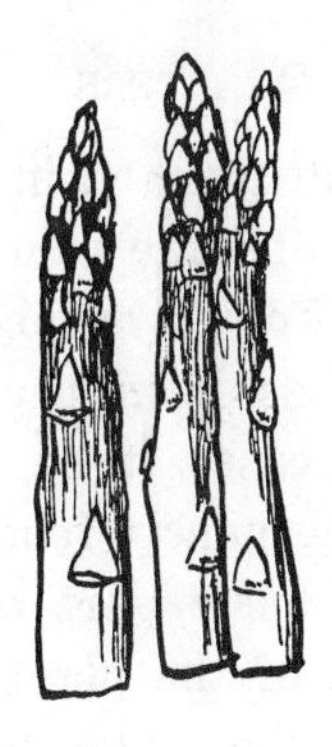

The way to find asparagus in late spring is to look for the old dried fronds from the previous year, which usually persist through the winter. This sport cannot be indulged in at a fifty-mile an hour clip but at ten-miles per hour, along almost any roadside in the spring, enough asparagus can be found to provide a good meal. Such hunting of asparagus has become a minor hobby with the author.

It is interesting to note that asparagus is one of the vegetables which was mentioned by the Greek and Roman writers. Cato, about 200 B.C., knew asparagus very well, and gave excellent directions for its culture. It was Pliny who first noted the fact that the wild asparagus in some sections was of excellent quality, while another Roman writer reported that Emperor Augustus was extremely partial to this vegetable.

Try hunting your own wild asparagus and see how good it tastes.

Chenopodium album — LAMB'S QUARTERS

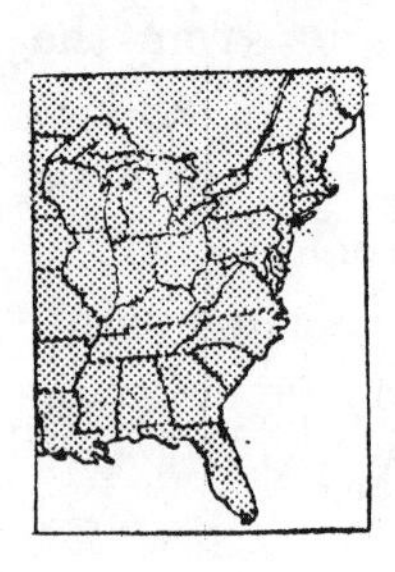

In many cultures and in many times, the seed of lamb's quarters (which is also known as pigweed or goosefoot) has been used for food. In some cases it has been used as an emergency food and in others as a basic part of the diet. Remnants of the plant have been found in the Lake Village

culture of Switzerland, which indicates that the seed has been used for food for untold centuries. This species of *Chenopodium* is, in fact, a double usage plant in that the tops may be used as greens when young, while the seed in autumn and early winter provide a "grain" nourishing enough to be used as bread. Because of the fact that the plants hold this seed until well after the heavy frosts, it has thus had value as an emergency food in times of famine.

The American Indians all the way from Mexico to the more northern climes have used one or another form of *Chenopodium* for bread-making. The dried or slightly cooked seeds are ground fine into a flour which tastes something like buckwheat and, because of the black color of the seed coat, has a similar dark appearance. The differences in the related species of lamb's quarters are largely of interest to the botanist. To the agriculturist this plant is just a nuisance weed, and it is doubtful if anyone will object to the harvesting of the seeds of lamb's quarters or to using the young tops as greens.

Capsella bursa-pastoris — SHEPHERD'S PURSE

There are a number of statements made as to the origin of the name of this undesirable weed, but in any case, the Latin name, translated, provides its common name. It is not a native weed, but one that has followed man everywhere throughout the world. Certain it is, that in shape, the plant's seed-pouches resemble an old-fashioned change purse.

Although it has been said that the tender leaves may be used as greens, and that birds are fond of the seeds, the greatest use of the plant has been for medicinal purposes, and in this field high values are claimed for it. The plant seems to have two properties, one as a diuretic, and as a haemostatic. Another use for this otherwise pestiferous weed is in the suggestion that the seeds be added to vegetable soup as a seasoning.

A mixture of the seeds of *Capella* with the seeds of *Chenopodium* provides a suitable food for many kinds of caged birds. In Shepherd's Purse we have a good example of how an otherwise despised weed may be very useful if one will but investigate the potentialities.

Asarum canadense — WILD GINGER

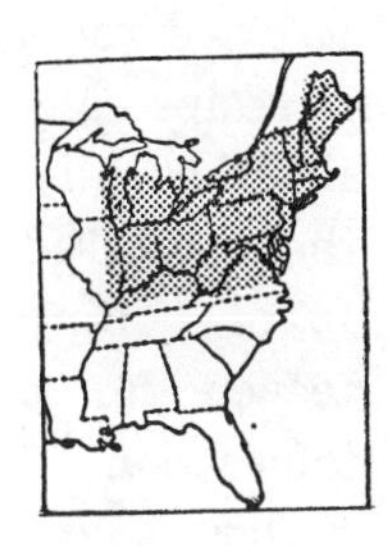

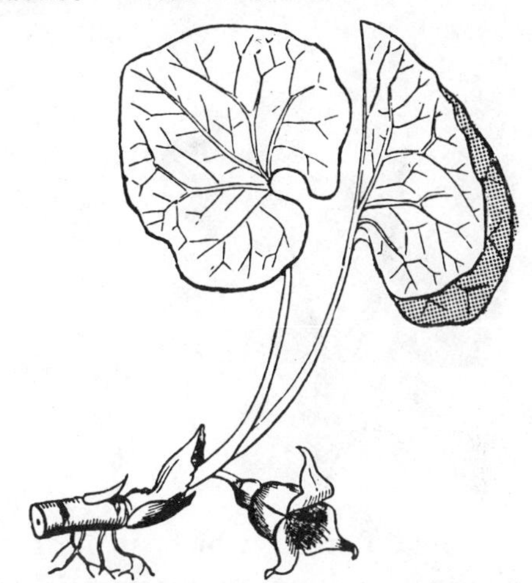

Not too commonly found in the woods, and then often difficult to distinguish because of its habit of living well hidden, is the plant which we know as wild ginger. The shape of the leaves is very distinctive (see illustration) and makes identification easy. The rather thickish root stock is fragrant, and may be used in either the fresh or dried state as a substitute for ginger. Gathered roots may be prepared for use as a candy, by cooking until tender, and boiling in a rich sugar syrup until candied, as one would prepare citron or grapefruit peel.

Coptis trifolia — GOLD THREAD

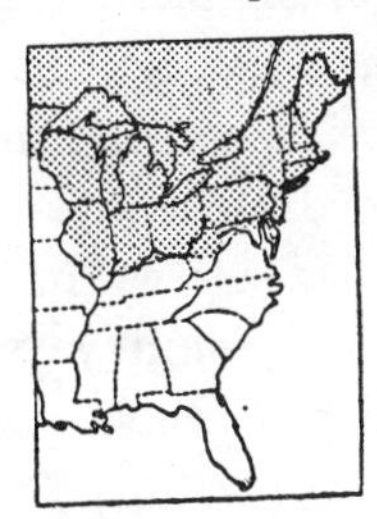

This is a plant which is still collected rather widely and sold for use by pharmacists. Its other common names are canker root and mouth root, both indicating its value in medicine. It is listed as being

an ingredient of a bitter tonic, and a mouth wash which is particularly effective in the case of ulcers. The entire plant can be used, but the dried roots are preferable. One recipe calls for the using of about a third of a teaspoon of the powdered dry roots to a cup of boiling water. This is steeped for a half hour and a tablespoon of the mixture is taken about every three hours.

Gold Thread, which, in its general appearance, somewhat resembles a strawberry, is of low growth, has shiny evergreen leaves, divided into three parts, which grow directly from the base of the plant. The small white star-shaped flower appears during the summer at the end of each flowering stalk. Further identification of the plant can be made by noting the bright golden-yellow roots which are much branched and frequently matted. The plant is happiest when growing in damp mossy woods or bogs, and is quite often found in association with sphagnum moss, or in dense shade among evergreens.

Cichorium intybus — CHICORY

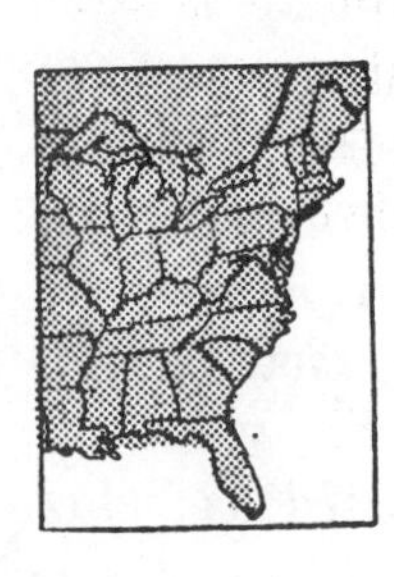

This is a very ancient plant that has come to America from Europe. Our most common association with chicory has involved the use of the dried roots of the plant as an additive to coffee in those southern parts of the country which have been touched by French culture. Northerners know it as Witlof Chicory.

The forcing of chicory roots to produce a blanched head of leaves has been a very large industry in Belgium, with imports still being made to the U.S.A. in spite of local supplies.

In our country chicory is a widely spread weed, which is best known in the northeast, for its often much-admired blue flowers. These flowers not only signal the location of the hard-to-exterminate weed, but, where plentiful, attract flocks of wild canaries.

Chicory is a close relative of the dandelion which it resembles in early growth, although it is much more vigorous where conditions are favorable. The perennial roots often become extremely large with age, and are frequently so long and well fastened in the ground that it is almost impossible to dig them out completely. It is just that same substantial root, however, which sends up a sweet and succulent growth in the spring, which, if gathered early, will provide edible "greens" of spicy and healthy properties.

As one ancient son of Italy said to the author when he was gathering it on a cold March day — "a little pieca da meat — this little salada, and a glass of wine — is good and it don'd a costa anything." Would that more of us Americans knew how to enjoy these simple things, as do such peasant-born inheritors of age-old knowledge!

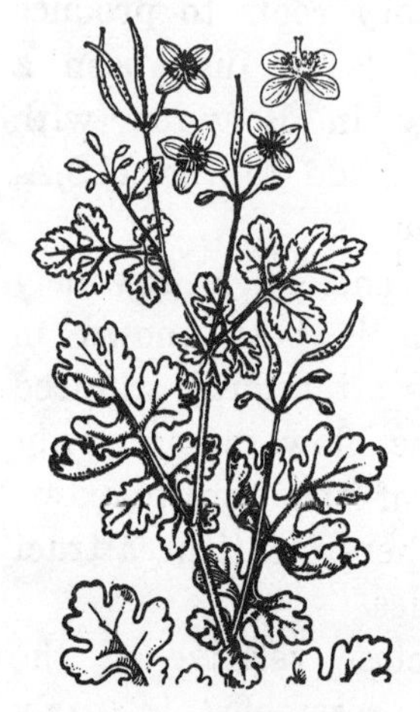

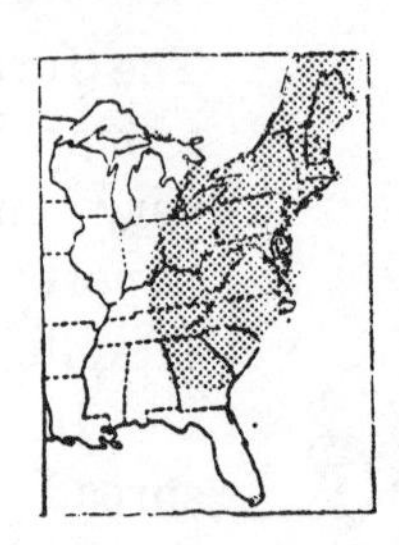

Chelidonium majus — GREATER CELANDINE

The easiest way to identify this plant is by the orange juice which is readily exuded from a broken stem, as well as from the root. Some of the other common names given to the plant, such as Swallow-wort, Tetterwort, Fellonwort, Wortweed, Killwort, Devil's Milk, all indicate the probability that the plant has been much used for medicinal purposes. The juice of the plant was once considered a certain cure for warts, corns, pimples, boils, and the like. Perhaps the curative values of this juice for warts is based on a phychological premise, but in any case, it is a harmless remedy to try, and a good plant to know about. The use of the orange juice as a dye is discussed in the appropriate chapter.

Equisetum fluviatile — HORSETAIL

One of the better names for this plant is scrub-grass or joint-grass. The latter name describes its appearance and the former its usage. Much of interest could be told about the equisetums, of which a number of species may be seen in the springtime.

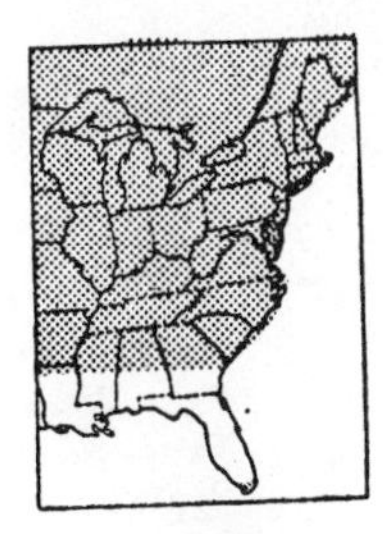

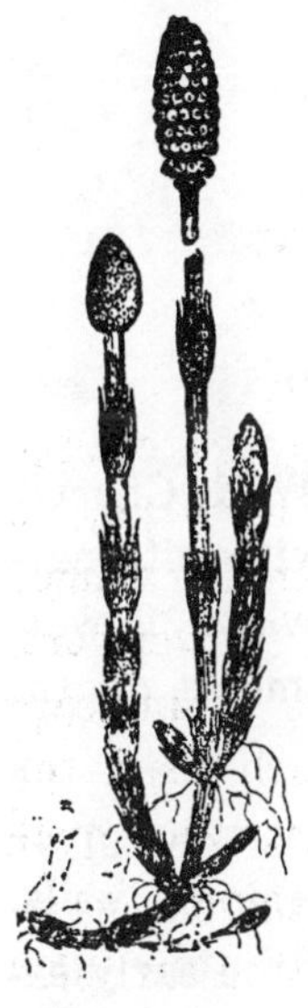

The equisetums are primitive plants, having been on the earth according to the best authorities, for an estimated three hundred million years. Botanists are generally agreed that the species of horsetail that now survive in our swamps were, in earlier forms, the forest trees of the coal age, often growing to a height of fifty feet. The name, equisetum, comes down to us from early times, because it was said that the Romans not only ate these plants for the starch which they contain but gave them their name, after the resemblance to a properly-tied horsetail.

Beyond the fact that they might be considered an emergency food, the commonest use which comes down to us from the Indians was the gathering of a number of the stiff stems together and using them as a scouring brush. The stems have a high mineral content, with a slightly abrasive character, and although science has given us plenty of modern substitutes for scrubbing, the nature enthusiasts might well like to try them in the kitchen, or for emergency camp use for utensil cleaning.

Cryptotaenia canadensis — HONEWORT

Another name for this plant is the Wild Chervil. The Honewort grows about two feet tall, with alternate and long-petioled three-part leaves. The white flowers are very tiny and are borne in irregular umbels on the top of the plant.

Generally throughout our territory, it is useful in the late spring or summer as a potherb. As the one name would indicate, it has a distinctive flavor, and the early explorer, Peter Kalm, reported in 1749 that it was being used by the French in soup. It may well be that it has a certain value as a source of vitamins and minerals. In Japan today the plant is cultivated as a garden vegetable.

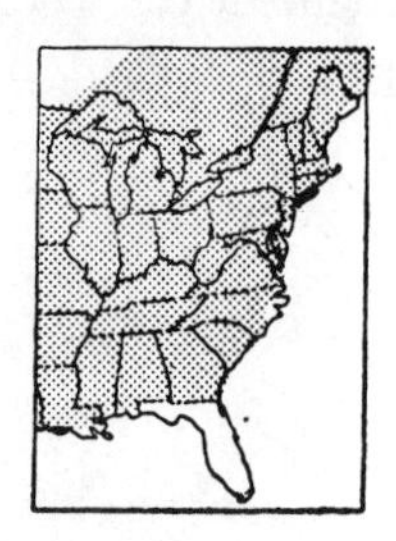

Eupatorium perfoliatum — BONESET

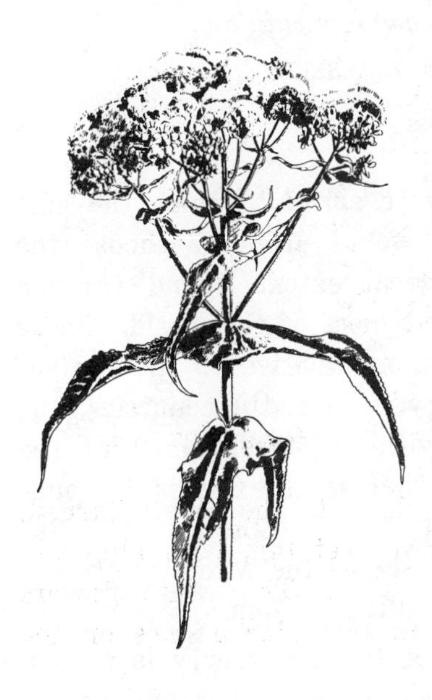

So many of the plants in America reveal in their common name the likely uses for the plant. Take for instance Boneset, which has no uses other than medical. In grandfather's day the parts of the plant used were the flowering tips, and also the leaves; an infusion of boneset being prescribed to alleviate the pain caused by broken bones as well as for similar pains from malaria and influenza. One old authority says that it is a "boon in miasmic districts, along the river, marshes, and so forth, and in all conditions where there is a great deal of bone pain." This is an interesting statement in itself because the Eupatorium grows only in wet circumstances, and can in such places usually be found without much difficulty throughout our entire region.

Another name for this plant, for which the reason is not so plain, is "thoroughwort."

Fragaria virginiana
— STRAWBERRY

If the average person were asked to name the best known wild edible plant he would probably choose the strawberry, which is, to a great extent, found (in one species or another) throughout most of the world. There are many interesting suggestions relative to the derivation of our name "strawberry." In Latin countries, the name usually is concerned with the fresh smell of the berries (French "fraises,") whereas the Germanic countries speak of it as an earth-berry — "Erdbeere." But there is nothing quite so delicious as the wild strawberry, and even more so when it is made into jam.

Although no description of the strawberry is needed to identify it, it would be well to note that one should always look for strawberries growing on the north side of east and west roads, for it seems to be most happy when the sun can strike directly from the south. Often also it will be found in open pastures or in lightly wooded sections, and always the strawberry seems to be at its best where plants can grow undisturbed.

Galium verum — YELLOW BEDSTRAW

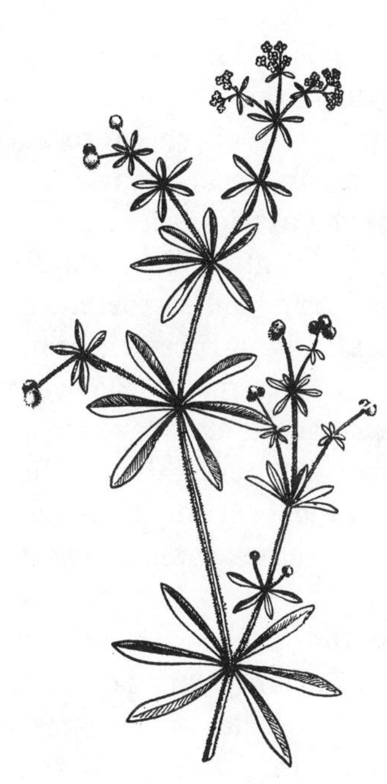

Another common name for this plant is the *cheese-rennet*. Both of these names are interesting in that they denote some actual or mythological use of the plant. A very pretty story is that this was the plant which was used to line the manger at Bethlehem when our Lord was born, and some books list the common name as "Our Lady's Bedstraw." This story might have arisen because the plant gives out a very pleasing aroma when it is dry, and in some countries it was much used as a filling for mattresses.

In times past it was also used to mix with rennet in the preparation of cheese, and the great botanist, John Ray, suggested that a refreshing beverage could be made by distilling the flowering tops of the plant.

Be that as it may, it is a lovely graceful plant when found growing with its feathery foliage and soft yellow flowers. For this daintiness and by reason of the stories connected with it, it is a plant that we all should get to know.

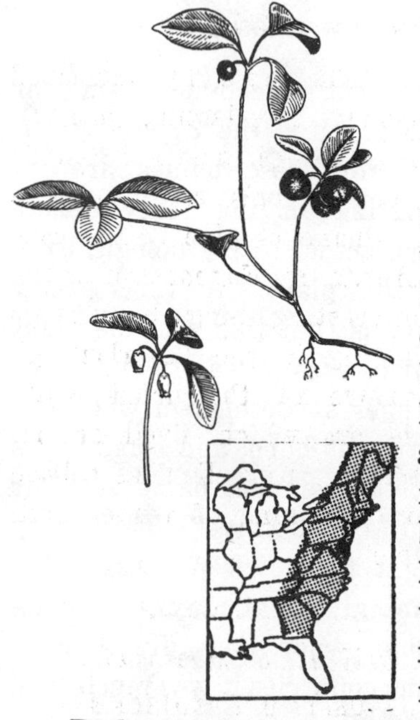

Gaultheria procumbens

CHECKERBERRY

As is true with its usual neighbor the bearberry, the wintergreen or checkerberry has many local names, including Partridge-berry, Boxberry, Mountain Tea, Chinks, Deer-berry, and so forth. The aromatic wintergreen is easily found throughout the year because of its dark evergreen leaves and delightfully flavored red berries. It is useful not only as an aromatic nibble, but also in medicine.

Perhaps most delightful is the chance to refresh oneself in the springtime with the tender young leaves of checkerberry, or nibble the aromatic berries when one is wandering through the woods in winter. It is said that many woodsmen use the mature leaves as a substitute for tea. In the 18th century, after the Boston Tea Party, this was especially true in the New England area.

Although there is some possible use for oil of wintergreen, in actual practice the commercial supply is now made by the distillation of the twigs of the black birch which provides the same flavor. The true oil of wintergreen distilled from the leaves of Gaultheria (when added to honey) is known to make a very soothing remedy for coughs and sore throats.

Gnaphalium obtusifolium — EVERLASTING

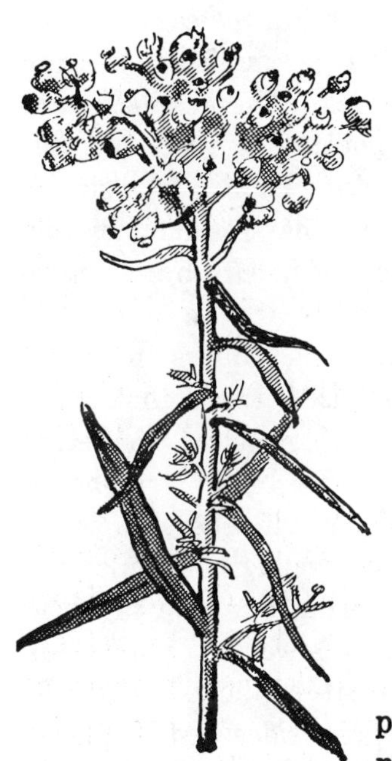

The easily identified Pearly Everlasting has no food values, but is listed in some herbals as a suitable medicine to give in the case of catarrh. For most of us the interesting thing about Everlasting is the "everlasting" nature of the dried white flowers which, dyed or natural, are excellent as a base or beginning of winter bouquets.

Helianthus tuberosus
JERUSALEM ARTICHOKE

In view of the fact that this perennial species of sunflower is native to central North America, it is interesting that it should bear the name of Jerusalem, but botanical philologists explain this by the fact that this English name is a corruption of the French "girosol," meaning sunflower. There is good evidence to indicate that this plant was cultivated by the Indians, who in turn introduced it to the Europeans.

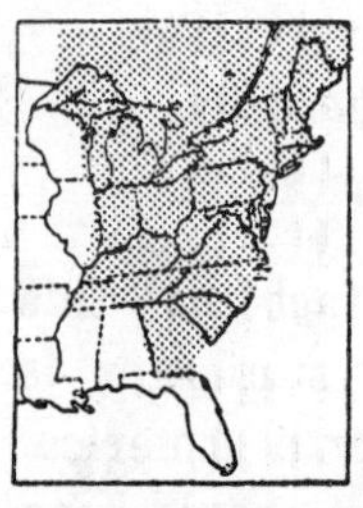

The Jerusalem artichoke is a plant which takes to cultivation quite easily, and the edible tubers are to be found in the early winter in most large markets of our eastern cities. These tubers are enlarged potato-like roots but unlike the potato have a low starch content. For that reason they are often recommended by doctors for those who are on a starch-free diet. The tubers may be cooked, or used fresh in salads, or pickled. They are even better escalloped with bread crumbs, a method which will absorb a large part of the mucilaginous juice.

Quite often one will find the tall yellow flowering plants of the perennial sunflower naturalized along roadsides and in places where there have been old gardens, showing that the plant is very persistent. Taking observations of the location of such easily identified plants on summer drives will provide an opportunity to return in the fall to dig the tubers.

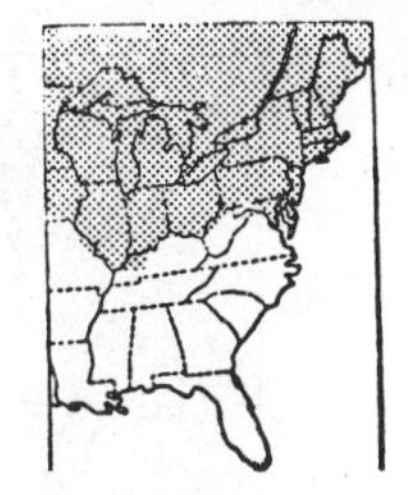

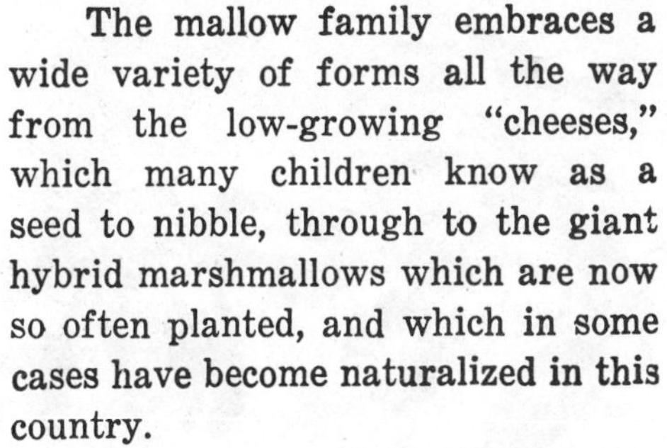

The mallow family embraces a wide variety of forms all the way from the low-growing "cheeses," which many children know as a seed to nibble, through to the giant hybrid marshmallows which are now so often planted, and which in some cases have become naturalized in this country.

The low-growing *Malva Sylvestris* (and other similar species) is a biennial plant which is listed in some herbals as a demulcent, which means that it is good for coughs. To use it for such purpose it is recommended that the dried leaves be made into an infusion with a little lemon juice and drunk when cold. The quality which has this medicinal value is found in the very mucilaginous juices of the plant.

Hybridized forms of the larger-growing native swamp Rose-mallows *Hibiscus Moscheutos* are among our most showy garden plants. Generally most species are not happy unless they can grow in places where they can have perpetually wet feet.

Phytolacca americana — POKEWEED

This is one of those common American plants which has had so many uses by so many people that it is known under many different names, some of which are *pigeon berry, garget, pocan, ink berry, cancer-jalap, redweed, pokeberry.* For all of these names there are good reasons, for the plant is at once a valuable edible plant, as well as an extremely poisonous one, and a description of it might well be included under both categories.

The poisonous part of the plant is the fleshy root, as well as the leaves and foliage when they have turned red; yet in the spring when the young shoots are about four or six inches above the ground, the tender solid stalks are one of the best substitutes for (if indeed not even a better plant than) asparagus. See illustration of shoots in Chapter 2.

The young shoots and the tender leaves of pokeberry can be prepared for the table by boiling in two waters until they are tender, and serving with a cream sauce. This is, indeed, a standard vegetable in the better gardens of the south, but throughout the northeast it is more commonly found growing in the wild.

Because of the extremely fleshy roots which gives the pokeberry a true perennial and persistent quality, strong plants may be located in the fall, dug up after the first early frost, and put into a warm cellar, covered with earth, and watered. Then, with sufficient heat, in time will come a succession of delicate pale green shoots which will be available for harvesting throughout the winter as a very tender and delectable green.

As might be expected from some of the names which have been given to it, the pokeberry also has uses as ink and for dyeing. The red fruits for use in dyeing should be picked only when completely ripe, care being taken not to leave them too long, because when too fully ripe, they are highly attractive to birds, who quickly strip the bushes. This same juice may also be used as a coloring for frostings in candies, note being taken that it is the actual seeds themselves which are poisonous, and not the juice.

A hundred years ago, the medicinal values of pokeweed were thoroughly explored and some values determined but in view of the poisonous character of the fruits and roots, caution is recommended for any use except the young shoots and the berry pulp for coloring.

Portulaca oleracea — PURSLANE

The common name of this plant has been perverted to "pusley," and the plant is so prevalent and weedy that it is usually ignored and despised. The succulent reddish-purple stems and small fat leaves are familiar to anyone who has ever had a garden, but there are many people, especially of foreign birth, who realize that these young leaves and stems, when cooked for about fifteen minutes in boiling salted water, are far better than spinach or other greens.

This "Pigweed" as the author was taught to call it, (see *Chenopodium* for true pigweed) does however cook

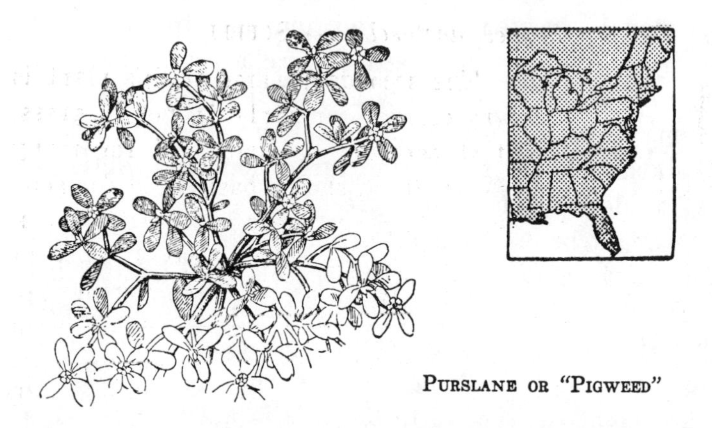

PURSLANE OR "PIGWEED"

down at an appalling rate, and it requires rather a large quantity of fresh greens to produce a dish for the table. The harvesting of quantities of purslane brings no tears to the gardener's eyes and in fact, the more used, the better. Another way of preparing this first cousin of the more beautiful garden portulaca is to fry it. In this case prepare diced strips of cold pork, partially fry them, add the purslane and cook down, and serve with butter, salt and pepper.

Certainly it may well be said that the old adage, "one can always find something good in the worst of things," holds true with the edible portulaca, as its Latin specific name tells us.

Rumex acetosella — SORREL

The secondary name of this plant is *sourgrass.* It certainly is not a grass, but it certainly *is* sour, and for many children it satisfies a basic need for sour things. It is an easily-identified low annual plant, with narrow leaves, lobed at the base like an arrow. It is common in gardens *everywhere,* and it has been known for centuries in Europe as a pot herb, although growing as a weed with us. In England in very early times this plant was cultivated for its leaves, just as spinach is today, and was held in great repute in the time of Henry VIII. As with any plant so common, it has a variety of folk names, most all of them being variants of the name sorrel or sour.

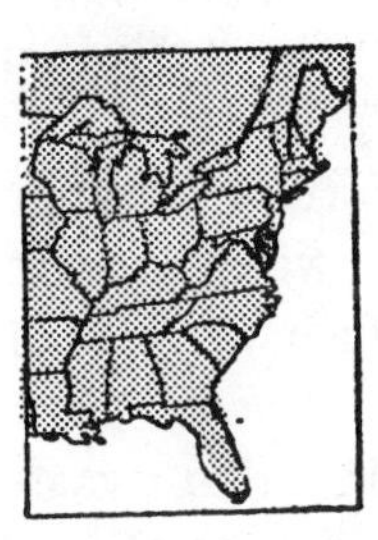

Those great cooks, the French, knew the particular values of this acid plant as an addition to the diet, and their "soupe aux herves" contains sorrel as a principal ingredient.

It is said that French cooks also procure from the sorrel a red coloring matter which is added as a harmless improvement to barley water, thus making for invalids an inviting drink which looks more like the red wine to which the French are accustomed. Here we see the culinary-wise French taking account of the value of eye appeal of food so especially necessary when one is sick.

Recipes for the use of sorrel in various dishes are given in the appropriate section.

Rumex crispus — NARROW-LEAVED DOCK

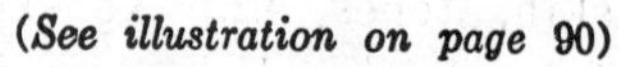
(*See illustration on page* 90)

Quite different in many ways from their close relative the sorrel (*R. acetosella*) are the docks, of which the narrow-leaved dock of this discussion is probably the most useful member. It is a persistent perennial herb, with lance-shaped curly-edged leaves. The root is long and thick, and literally defies any attempt to pull it up.

The late-lamented scientist, A. C. Kinsey of "sexual" fame, in his book on edible wild plants indicates that practically all of the fifteen species of docks are useful as wholesome greens when young. But from the writer's own experience, the narrow-leaved dock is superior, not only among docks but also to spinach and other cultivated greens.

The dock is a good example of one of those strange situations where a particularly persistent weed is also an interesting and useful one. The most lush growth of the weedy docks will be found growing in ash piles, rubbish heaps, along rich roadsides, and in other waste places where the roots can search widely for food and remain undisturbed for long periods. The young leaves should be used when a foot or less in length, cooked like spinach until tender, and served with proper seasoning and butter.

In addition to its value as food, *R. crispus* is listed as one of the more important medicinal plants by the Department of Agriculture. The part used is the root, which is collected late in the summer, after the protruding tops have turned brown. The roots are washed and then split into two parts and dried. All authorities agree that an active principle in this dried root has values in curing infections of the skin, itches, etc. An extraction has also been used widely for relieving liver complaints. In simple practice, it is said that the freshly-gathered leaves, when well washed, may be laid on sores as a healing agent. The same principle which is active in the narrow-leaved dock is also found in other forms or species of the *rumex*.

Beyond these more practical uses, the garden club ladies have found an important use for dock in the interesting form of the dried seedheads, with their warm red-brown color which are everywhere a prominent feature of the fall landscape. Additionally there is some record of the dried seeds having been used as a cereal grain, and what with one use or another, we find that this often despised weed of the roadside has much to offer for the exploring householder.

It is perhaps difficult at times to bring ourselves to make much use of weeds, because of our negative conditioning to the word *Weed* itself. The nature lover must often combat this tendency to "classify" plants, and we see in dock a good example of this difficulty. Such an attitude towards "weeds" has been true for centuries, for the noted herbalist Culpepper, in 1653, had this to say about the dock plant of this discussion:

> "All docks are boiled with meat and are as wholesome a pot herb as any growing in a garden; yet such is the nicety of our times, forsooth, that women will not put it into a pot, because it makes the pottage black; pride and ignorance (a couple of monsters in the creation) preferring nicety before health."

Saponaria officinalis — SOAPWORT

A more happy name for this plant is that of Bouncing Bet, and as a weed in the garden bounce it does. It is really a beautiful plant, with flowers not unsuitable for cutting, belonging as it does to the family of pinks. The name of soapwort comes from the fact that it contains an active element called saponin, which produces a soapy froth in high concentration in the roots.

In medieval times it is probable that mendicant monks used the leaves as a substitute for soap in washing their clothes and perhaps a knowledge of the soapy qualities of *Saponaria* might well be useful in any period of soap shortage.

THE GOLDENRODS

No book on American plants could ignore the existence nor indeed the importance of the many, many forms of goldenrod which are so characteristic of our country, two species of which are pictured opposite. To say that the goldenrod has small use would be to ignore its great value in bouquets, and certainly it is only its weedy nature and commonness which keeps it out of our gardens. To the English gardener, this is one of the choice autumn garden flowers, and the American gardener might be surprised at what selected plants of the goldenrod would do for him if transplanted to a fertile spot in his garden.

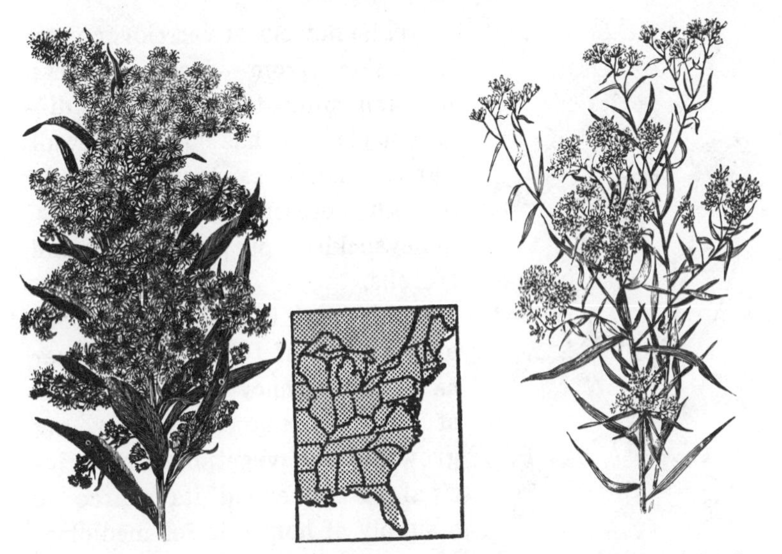

Solidago odora, a particular species which should be looked for is not pictured, but it can be identified by the delicate anise-like odor which the bruised foliage gives off when crushed. The flower head is a one-sided one, the leaves are about three inches in length, very smooth and wide-spreading. The dried leaves of this sweet goldenrod were used during the Revolution as a substitute for tea and probably could be so used now if needed.

Trifolium pratense — Red Clover

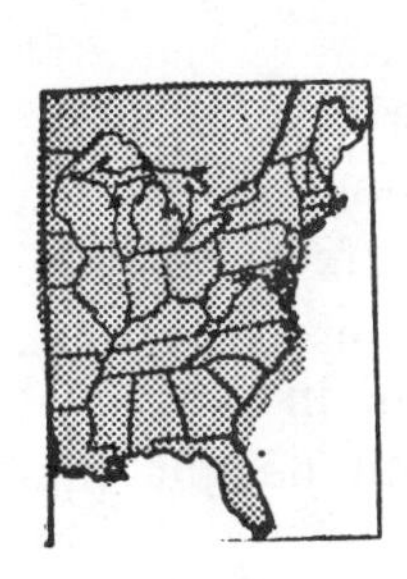

One of the most beautiful native wild flowers of pastures and meadows is the red clover. The value of clover is well known to the bees, and certainly there is no honey which is more delicately flavored nor more beautiful than clover honey.

The flowers of red clover vary in color from deep magenta through pink to white. In different places it has borne different names, such as (appropriately enough) bee-bread, cock's heads, honeysuckles, purple clover, and trefoils.

Although there is some record of the value of the seeds of clover as an emergency food and even of the cooking of the young growths as a vegetable, the principal value beyond its source as a supply of honey, is for medicinal purposes.

Tea prepared from the dried flowers of clover has been given for many years as a soothing medicine for people suffering from bronchitis, asthma, and coughs, while for whooping cough, it has especially been recommended. As a food forage crop it is, of course, well known to the farmer, but these other values are perhaps less well known.

Taraxacum officinale — DANDELION

"How like a prodigal does Nature seem,
When thou with all thy gold so common art."

Thus aptly does the poet sum up the several sides to the dandelion. One of the most widely-spread weeds in the world it is nevertheless a plant which from earliest times has been known to have many useful properties. The particular householder may well curse the ineradicable dandelion in the summer, but will sip with delight the golden brew that is dandelion wine.

Actually the only bad feature of the dandelion is its persistence, a quality well praised in most plants. The name of dandelion appears among the food plants, in medicine, and who shall say that not the least of its values is found in the beauty and wonder that the growing child finds in the seed heads, as they are used to tell fortunes or the time of day.

Evidence of the early medicinal use of the properties to be found in the roots of dandelion are shown by its once Greek (now Latinized) name: *Taraxacum officinale*: for the Greek word *Taraxos* meant *disorder* and *Akos* remedy: while the *"officinale"* or specific name indicates its place in the official list of medicinal plants.

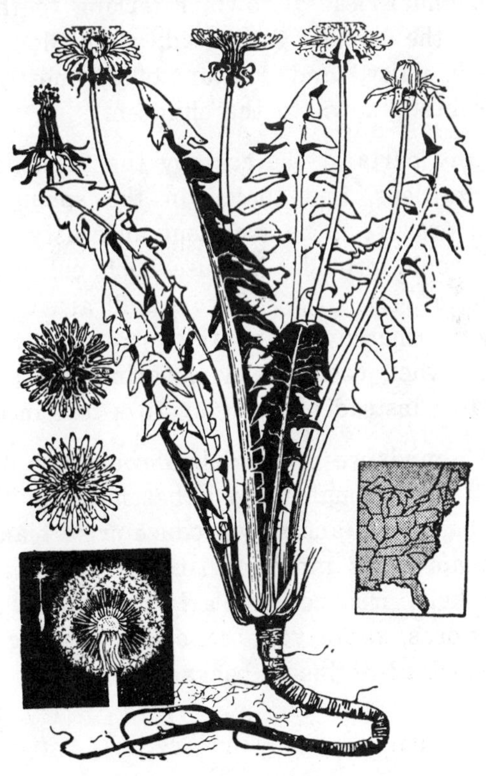

The natural desire of earlier generations, much starved for "greens" through long winters, to eat the newly sprouted greens had a well-founded medical history, although now when "greens" are available all winter, we may not feel this urge. Urge or not, it is good to know that "eat your spinach" is not just a fad, and that the weeds on the lawn may well be even more healthful than spinach.

Looking to common names for other interesting ideas we find that the leaves of dandelion are well described in the "dents de lion" (French for "Lion-toothed") which we have perverted to dandelion. Among other common names we find Priest's Crown, referring to the bald appearance of the denuded seed head, while Blow-ball, Timetable, Irish daisy, and Fortune-teller mostly refer to properties best known to the children.

In many parts of the country the dandelion is cultivated in gardens, and forced in the spring with cold frames, to procure more succulent and lusty greens than the lawn would provide, because such cultivated dandelions have a less strong flavor and are more tender. From lawn-gathered greens this strong flavor may be eliminated, when cooking, by covering the greens with boiling water instead of bringing to a boil in cold water.

If an abundance of dandelion roots can be gathered in the fall and brought into a basement in the winter, they may be forced, and will produce crisp blanched salad leaves, by no means inferior to the forced witlof chicory of the markets, produced similarly from roots of chicory. In other words, make your dandelion digging pay some dividends instead of just a backache.

Urtica species — THE NETTLES

Whoever has been stung by nettles will recognize the name, and any recommendation of the use of the plant for culinary purposes should be accompanied by a warning to put on gloves when picking them. Nettles are widespread and disagreeable weeds and any gardener will be glad to cooperate by giving his supply to those who wish to experiment with them.

But, in some countries, especially in Scotland, the nettle has been much valued for use as a pot herb in the spring when the tops are tender.

As far as other culinary uses are concerned, there seems to be a certain coagulant quality in nettle juice which in some places in Europe has been used as a rennet to thicken milk. This special quality may explain the inclusion of nettles as a constituent of mixed herb dishes, such as is recommended in the recipe section.

Medicinally, in herbal practice, nettle tea was used freely as a blood purifer and as a possible cure for rheumatism. Another interesting use is as a fiber for textile purposes. In Britain nettle has been used as a substitute for flax and it is said as well, that an excellent paper may be made from these fibers. Howsomever, as with so many other weeds, there are many values to be discerned if one will literally get beneath the stinging surface and study a plant such as a nettle.

Verbascum Thaspsus — GREAT MULLEIN

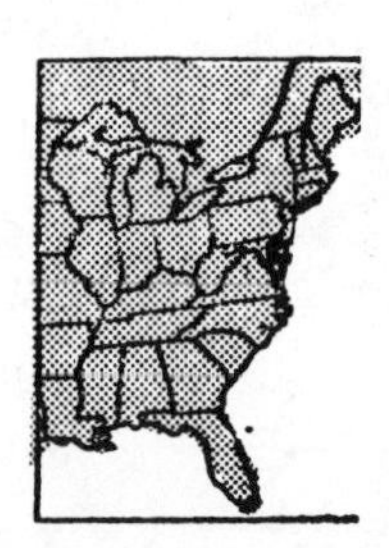

Because this plant is so widely spread throughout the countryside, it has achieved various local common names. In addition to "mullein" it is known as: velvet dock, Aaron's rod, Adam's flannel, blanket leaf, bullock's lungwort, candlewick, feltwort, hare's-beard, and velvet plant.

Considerable use of the mullein has been made in late years by the garden club ladies, who have recognized the inherent beauty in the rosettes of velvety gray leaves which are especially beautiful when the plant is coming to life in early spring.

Dyers have found in mullein the source for a bright yellow dye, while the "medicine men" rate a decoction of the dried leaves as very valuable for irritated throats. Certain it is that the landscape value of the plant is considerable, even though it is a weed, and in any stage, from its rosette form in spring, through the flowering stage in summer, to the dried warm brown seed stalk, it is a plant with many possibilities for artistic persons.

USEFUL AND BEAUTIFUL FERNS

Everywhere in the eastern United States spring is heralded by the unfolding fronds of ferns. The family of ferns is a highly variable one in many ways and among the species are a few that should be noted here.

Through the dry open woods and over recently burned clearings and pastures, one finds growing a rather coarse fern with its scattered stalks and tall fronds. This is *Pteridium aquilinum,* or Bracken. Related species of Bracken (or Brake) are common from Newfoundland to Mexico. In the spring, before the young fronds are fully uncurled, they are ideal as a spring green. To people who reside in a city such as Boston, or up in Maine, the use of these croziers is common practice as they are usually for sale in the big markets.

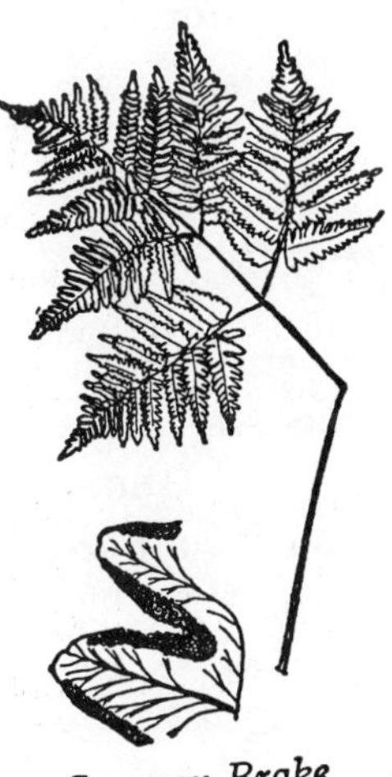

Common Brake

These stout unfolding stalks can be picked when they are about six to eight inches high. Draw them through the hands to remove the wool-like covering; then wash and bunch like asparagus. Preparation is by boiling in salted water, or steaming until tender, seasoning with salt and pepper and with butter added.

Ostrich Fern

The Ostrich fern, *Pteretis pensylvanica,* also provides a very edible new growth. This fern can usually be identified by the way in which the new fronds (covered with a brown papery scale) arise out of a vase-like clump. This fern is found in the rich beds of old streams through southern Canada all the way down to Virginia. The unfolding fronds can be picked just before they are completely uncurled and they, like the Bracken, provide a very desirable substitute for asparagus.

Another genus, *Osmunda,* (not illustrated) provides us with, in *O. cinnamoma,* with what in Maine are called the true "fiddleheads," another edible crozier which is often canned and sold.

Again, the Regal fern, *Osmunda regalis,* which is usually found in peaty swamps and along streams and is quite a giant among ferns, is the source of Osmunda fiber, much in demand among orchid fanciers, professional and amateur. The fiber is really the wiry, black, root growth, which when freed from the mud in which it grows, is an ideal medium for the potting of orchids of almost every kind.

The name of "crozier" comes from the fact that the young unfolding stems resemble the staff or crozier of a Bishop. Looking closely at a crozier one is aware of the dynamically artistic form that nature has here produced, which reveals in its spiral shape, the same expanding curve that one finds in conch shells, in pine cones, or sunflower heads.

"He who knows what sweets and virtues are in the ground, the waters, the plants, the heavens, and how to come at these enchantments, is the rich and royal man."

EMERSON, *Essays* (Second Series): "Nature"

Chapter XI

LOOK FOR WATER AND THE PLANTS OF WET PLACES

Without mention of the Water's Edge no book on the highways of northeastern America would be complete, for in view of the many miles of coast line, the ins and outs of the shores of the Great Lakes, rivers large and small, plus innumerable lakes, ponds, and streams, water is very much in the landscape of the wayfarer. That there are fish in all these waters goes without saying, but as fish are not "still life" we shall leave discussion of fishing to the followers of Isaac Walton. But just as it is said that no botanist can properly botanize without getting his feet wet, so it should be noted that the swamp, the quiet lake edge, or the tidal shore offers many rewards of edible or otherwise useful things.

Of all the types of wet areas probably the swamp is most crowded with plant life. Here we find such useful plants as the sphagnum moss, the blueberries, elders, skunk cabbage, and others. Still waters of the shallow ponds yield rushes, iris, sweet flag, and cattails. On the salt water shores we find rewards of another kind, such as the useful sea weeds and the not-so-mobile forms of sea food such as clams and mussels.

Acorus Calamus — SWEET FLAG

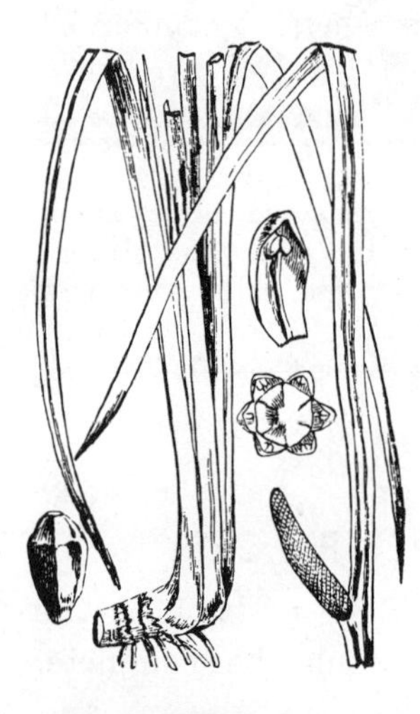

In addition to the name of Sweet Flag, this plant is known by such titles as Myrtle Sedge, Myrtle Grass, Sedge Grass and Beewort. In its characteristics it most resembles members of the Iris family, which grow in similar situations. But the Sweet Flag can be easily distinguished from the iris by the aromatic fragrance given off by its leaves, its yellowish green color, and more especially by the minute flowers borne on a finger-like spike, as shown in the illustration. The root is a rhizome which runs horizontally near the surface and is the useful part of the plant.

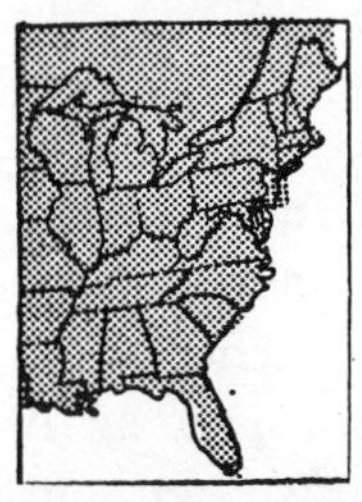

Growing in almost all parts of the temperate world in moist places, this plant has been used for centuries for medicinal purposes. It was a drug plant of the Greeks, and was certainly the panacea of the Indians. The roots may be gathered in the autumn or spring, and after washing, dried in a moderate heat. The unpeeled dry root may be chewed raw for stomach ailments, or used as directed in the chapter on medicinal plants.

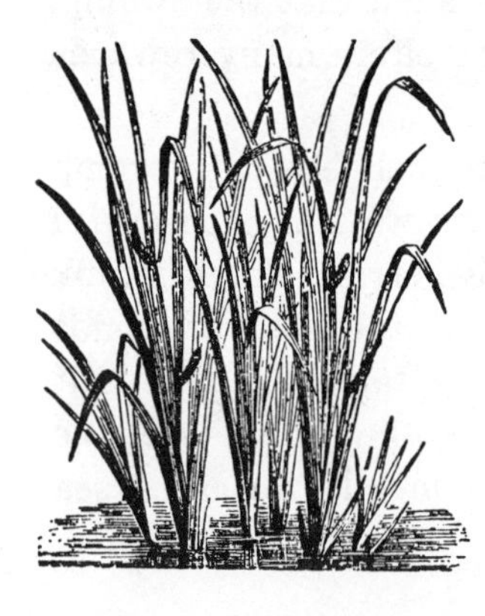

Iris versicolor — Blue Flag

This is the lovely member of the Iris family which is found growing along the edges of low ponds or even on the spring-fed edges of the seashore. Its blue flowers in spring are quite as beautiful in their way as the cultivated Iris, and the plant is surely one that should be known and admired by all.

From a useful standpoint, however, it has a variable reputation, as there is considerable evidence to show that the plant is actually poisonous, and special note should be made of the difference between this plant and the edible sweet flag, because the two plants grow in similar habitats, and have a superficial resemblance. As noted in the discussion of *Acorus*, the *sweet flag* has a yellowish green foliage, and gives off an aromatic fragrance, whereas the blue flag has a darker green foliage quite without fragrance.

The flag is discussed here because of this difference between the two species, and because of the fact that there has been some use made of the roots of blue flag for medicinal purposes. It is not recommended that any experiments be made along this line by the average person because of uncertain reactions.

Heracleum maximum — COW-PARSNIP

There are a number of forms of this giant parsnip which grow in rich meadows, moist thickets or seashores. The plant is variable in height (according to the location), running up sometimes as high as ten feet under good conditions. The young growth of the cow-parsnip or master-wort is said to be quite as palatable as stewed celery, having an aromatic and sweetish flavor. Its use cannot be recommended from experience, but there is considerable evidence that many Indian tribes used one or another parts of the plant in their diet. One observer states the fact that the strongly flavored basal part of the plant was used by the Indians as a substitute for salt.

Watch out, however, and do not confuse this with the somewhat similar (although smaller) poisonous members of the water hemlock family described in Chapter 13 under *poisonous plants.*

Impatiens pallida — JEWEL WEED

There are a number of interesting things about this plant which one often finds growing in very wet places on the edge of streams. Very characteristic is its exceedingly soft or succulent growth, the stems when broken exuding an orange juice. It seems to have been a wise provision of nature that the plant is often found growing near great masses of poison ivy, and there are a number of allusions in botanical literature to the fact that the juice of this plant will check the spread of ivy poisoning, at least as effectively as many other remedies.

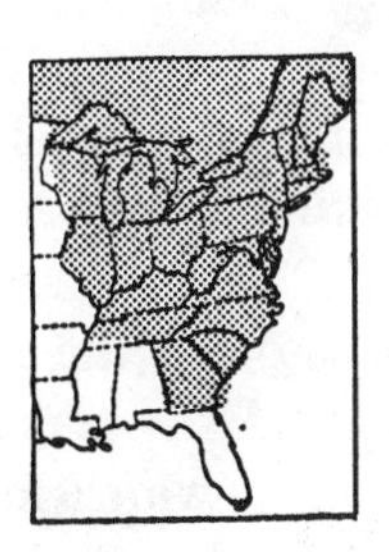

Another interesting fact about the plant is that the seeds when ripe disperse themselves through a mechanical method, the seeds often landing at some distance from the parent plant.

Names for the plant other than jewel weed include lady slipper, snapweed, lady's eardrops and touch-me-not. As there is some record that animals have been poisoned by eating the juicy plants, it is not recommended as a food plant. It is however, well worth a try, in an emergency, as an antidote for poison ivy.

Chondrus crispus — IRISH MOSS

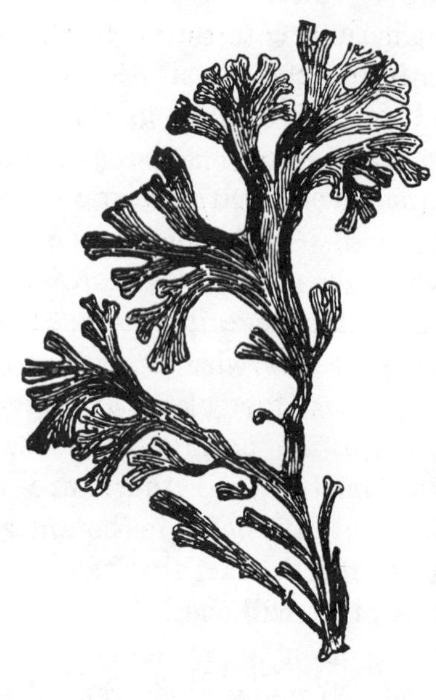

It is unfortunate that the name of this plant suggests that it is of foreign origin, for actually it is found growing all along the shores of the northeastern United States, right up from the Carolinas. It may be found at (or just below) high tide, growing chiefly upon rocks, but very often separated and cast up on the beach by high tides. It can be identified by the general shape (illustrated) and by the tough but elastic "fronds."

Irish moss when found in the fresh state may range in color from white, (naturally bleached) through cream and green to purple. The mucilaginous quality of Irish moss is the factor which makes possible its use in blancmange. The red color of the dried moss quite likely betrays the presence of iodine, thus making it a healthful food for everyone.

A number of herbals give various uses for the medicinal use of chondrus, especially because of the combination of iodine with the mucilaginous quality, which makes it an excellent remedy for chronic sore throat and similar conditions. Chondrus may also be the base for a hand lotion, one recipe suggesting that for this purpose it be added to the juice of the cucumber.

Complete instructions for its use may be found in the recipe section.

While we are discussing a "moss" which is not a moss at all, but a form of algae, it is interesting to notice how the name moss has been applied to a number of plants which are botanically quite unrelated to the true moss. In this book the only real moss considered is the peat-moss (sphagnum), but under Lichens mention is made of Reindeer-Moss. Then there is the quite commonly known club-moss, which is a *Lycopodium,* a primitive non-flowering plant which is more fern-like than mossy; and finally two different plants known as Spanish-moss: the California-grown kind, which is a lichen (*Ramalia*), and the very familiar Spanish-moss of the southern states (*Tillandsia*), really a tiny member of the same bromeliad family to which the pineapple belongs.

Pontederia cordata — PICKEREL-WEED

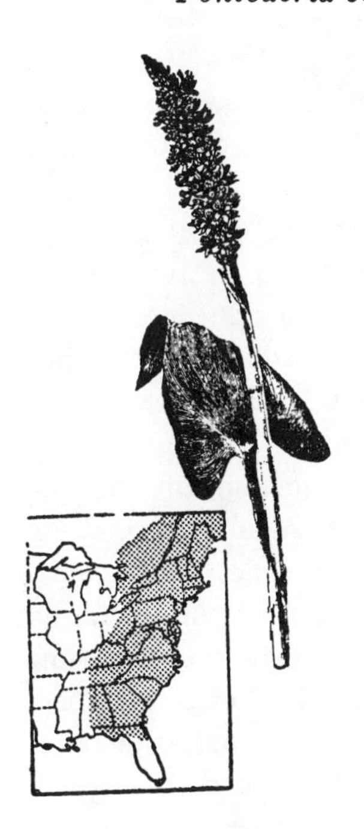

This soft-growing plant of the muddy edges of ponds has an arrow-shaped leaf, near the summit of which is borne the thick flowering stem topped with a spike of beautiful violet-blue flowers. These flowers are followed by a fruit containing a solid starchy seed and it is this seed which at times has been used as a grain. It may be gathered when tramping through the country in the early autumn, as a means of allaying one's hunger. Some authorities say also that the young leaves may be cooked as a succulent green herb in the spring.

Nasturtium officinale — WATERCRESS

There is a great body of literature which has developed around the use of this humble little "water-weed." Our forefathers deemed the plant to be only good as an appetizer, but more recently scientists have found that besides its snappy flavor, the watercress contains a quantity of valuable vitamins and other essential food elements. We know that the Persians, the Greeks, and other ancient herbalists knew or suspected the presence of valuable medicinal values, and recommended cress "to make children grow strong," and as a "remedy against scurvy." One sixteenth century writer went so far as to say that "the eating of watercress doth restore the wonted bloom to the cheeks of old-young ladies."

Although now found as a common wild plant throughout our territory, the watercress is not a native American plant, but has been naturalized from Europe. As in the cases of many other plants, the botanists are not entirely agreed on its Latin name: it is variously known as *Radicula* or *Nasturtium aquaticum.*

Returning to watercress, and ignoring any of the possible medicinal values about which so much has been written, we can say that this plant is one of the most desirable "finds" of the wayside. Its habitat will be found to be a cold running brook, and mid-spring the best time for gathering. Never pull the plant up if it can be helped, because this destroys the roots for future use, the basal roots being in any case quite useless for food.

Care should be taken not to use watercress which is growing where there is possible drainage from farmyards. In all cases it should be washed carefully before using to remove the dirt and possible contamination, *regardless* of where it is gathered.

Mentha (*species*) — THE MINT FAMILY

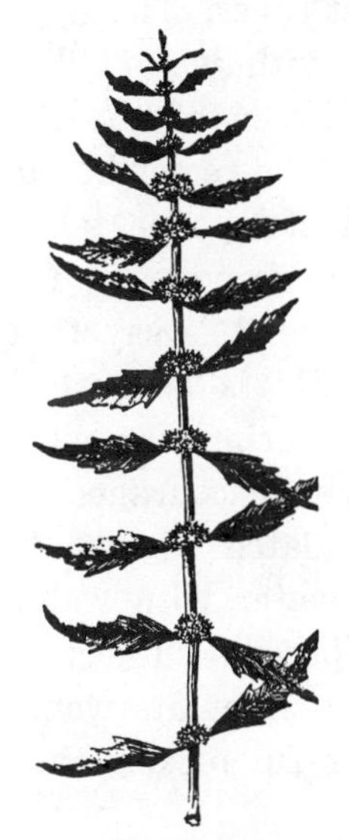

Within our region there are a number of forms of mints which, in one way or another, are useful and delightful. Many species are so closely allied that it is difficult for any but an experienced botanist to separate them. The mint family itself may always be determined by the novice through an observation of the stem, which instead of being round, is always exactly square, with the leaves arranged oppositely. Members of the family have a pungent aroma, variable in each species, but always present when the leaves are crushed.

The mint family furnishes our kitchens with a great number of the herbs which brighten our cooking: sage, thyme, marjoram, summer savory, as well as spearmint, peppermint, horehound, and others. The mint family is, in short, a very large and interesting one, easy to identify, and of varied uses. In our section, the commonest of all the mints is the spearmint, (*M. spicata*) which, having escaped from cultivation, is quite commonly found growing in or near slowly running streams and is familiar to all country children.

Nymphaea odorata — WATER LILY

To discuss the usefulness of the water lily is certainly to ignore its highest value as a decorative flower. Without doubt the fragrant water lily is one of the loveliest of flowers, possessing a beauty and delicate fragrance known and appreciated by man since the earliest times. The sight of a small pond filled with water lilies set against the glossy green of the lily pads is something long to be remembered.

In picking the flowers, care should always be taken not to pull up the entire plant, thus destroying it for future lovers of beauty. Wild flower authorities say that through thoughtless picking the water lily is being lost to many natural beauty spots. The indiscriminate picking of water lilies by children is cleverly discouraged in Holland, where there is a much circulated legend to the effect that any boy falling into the water with water lilies in his possession will immediately become subject to fits. One can easily understand how the fear of fits would tend to discourage any but the most careful picker, thus saving lilies as well as lives.

If one were disposed to explore the medicinal claims for the water lily, he would find that it is rated to be variously good for stimulation of the scalp to prevent baldness, as well as an external and internal medicine for almost everything from dysentery to freckles. Let us rather leave the water lilies blooming along the roadsides to give us more years of fragrance and beauty.

Symplocarpus foetidus — SKUNK CABBAGE

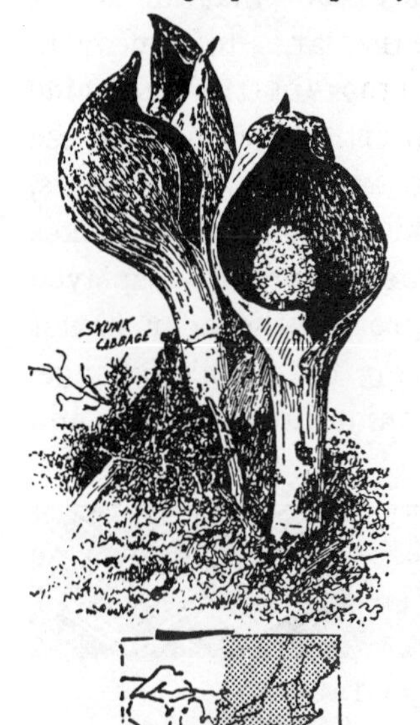

The American Indians — who called this plant "skota" — valued the skunk cabbage as an important medicine, and we in our pharmacopaea, under the name of *dracontium,* use it as an emetic, narcotic, and anti-spasmodic. Obviously with those qualities, a large dose of it would be poisonous, but the plant well advertises its dangerous nature by its fetid and objectionable smell. This rank odor, present in the lush foliage, is characteristic of deep swamps on almost any warm day in late spring.

In the very early spring growth, the leaves are found clasping each other tightly in a cabbage-like manner, and it is this tight tuft of leaves that makes a not unpalatable vegetable. No trace of the odor seems to be given off in cooking, but as with pokeberry shoots, it is suggested that the heads be cooked in several waters, to which a pinch of baking soda has been added.

In discussing its use as a medicinal plant, records indicate that the Indians used the powdered and ground-up root of skunk cabbage as a styptic to stop the flow of blood on surface wounds. They also flavored other medicines with it, to cover up even worse tastes or odors found in other plants. A powder made from the roots was prepared as a bread by the Iroquois Indians. In spite of its forbidding name, the plant seems to have a number of possible uses.

One may further point out in its favor that the skunk cabbage is one of the first flowering plants to appear in the spring, growing as it always does in the protected warm depths of swamps. The artistic person will see possibilities in the cluster of the emerging leaf heads, which with a few of the very interesting flowers, makes an unusual floral arrangement when properly displayed in a low bowl. At this stage of growth the odor is not offensive. Through using and knowing the skunk cabbage, one can learn that bad names may hide much beauty and usefulness.

Typha latifolia — CAT-TAILS

A useful plant from time immemorial, it is probable that some species of the cat-tails were the "bulrushes" in which Moses was hid as a baby. Certainly we do know that the long, broad, and tough leaves have been used throughout the ages for the weaving of matting, for chair seats and baskets, and that our own American Indians wove a covering for their summer wigwams from these leaves. For these practical purposes, the cat-tail can be rated as one of our very valuable American wild plants. Looking for all possible uses of

the plant, we need but refer to the common names by which it is known, such as flag, marsh beetle, flagtail, blackcap, water torch, candlewick, as well as some of the Indian names that have been translated as, "prairie chicken feather," "eye itch" and "roof grass."

The most important modern usage of the cat-tail has been for the seating of chairs, rush seating still being a native art for which there is considerable demand, and a skill which, with chair caning, is still being taught in schools for the blind. No chair seat actually is any more comfortable and long lasting than one made of the twisted and woven leaves of cat-tails.

During World War II when other products were in short supply, a Chicago scientist promoted the use of the water resistant floss (or down) of the ripe heads as a stuffing for toys and life preservers, and as a padding in tanks and airplanes. This was no new idea, for it is this same floss which was used by the Indians for padding their cradle boards, and for the making of exceptionally warm sleeping bags.

Beyond these practical uses, there is considerable record of the value of the plant for food, especially among certain Indian tribes. The part so used was the thick leading shoots of the root stock, which was, and still may be, eaten as a salad or cooked vegetable.

The most commonly appreciated quality of the cat-tail is its ornamental value. Cut and dried when the heads are just fully developed, they provide a bold note in the winter bouquets of dried wild plants, which are now, after a lapse of a half century, again becoming fashionable.

Sphagnum symbilolium — SPHAGNUM MOSS

Growing everywhere throughout the world, this beautiful moss makes up the principal plant growth on the bottom of many swamps, and if one dug down far enough he would find that the overlapping growths of this sphagnum go to depths of twenty feet, thirty feet, or even more. In fact, it is from this piling up of successive growths that we secure the material which is much advertised and so well known to all good gardeners as "peat moss."

A comparison of sphagnum with other mosses points up the great difference between the various forms of the mosses. Certainly the sphagnum can be said to be the most useful from every standpoint. This is so because sphagnum, either in the dried-live or dead state, has the property of absorbing twenty times its own weight in water. It is easy to see why, in garden use, the moss is so valuable to plant growth. The cells retain these great quantities of water over long dry periods, giving it up to thirsty plant roots only as needed.

It would probably be very difficult for the ordinary gardener or autoist to hope to secure any quantity of the deeply-buried peat moss, because of the difficulty of digging it out of the depths of wet swamps. But it is entirely possible to make many uses of the living sphagnum which, as the terminal growth, is easily secured from almost any road-side swamp.

As an emergency application to a wound to absorb blood, a handful of sphagnum moss with the water squeezed out would be extremely useful. Its ability to absorb moisture is twice as good as that of cotton, there also being good evidence of the moss's having an iodine content and healing acid properties. Considerable experimentation is now going on as to further medicinal values of the sphagnum, but in any case one can be certain that sphagnum is quite sterile and useful as an emergency bandage, especially when found in remote swamps.

More commonly known than this use, however, are its many values for the garden. Articles have appeared recently in gardening magazines relative to the use of the live or dead sphagnum (finely ground up) as the ideal medium for the germination of delicate seeds. Nurserymen for many years have put it around the roots of plants when shipping, and the florist has made it the base for stylized, solid, "pieces."

One nurseryman in Massachusetts has, however, gone much further than this in using pure live sphagnum as a complete growing medium for certain acid-loving plants. Plants so grown include species of the holly and many other ericaceous plants. These he has grown for a period of three years or more, right through from the cutting stage to a saleable plant, all without the use of any soil.

In this new method he uses a discarded one-quart lubricating-oil can as a pot which, after having been provided with a drainage hole at the bottom, is dip-painted. Filled with sphagnum moss, the rooted cutting is planted in it, and here it can be left for a long period. Even in the dryness of summer such a container needs but little attention except shading, for the occasional rains supply the needed water.

Experimentation has proven that there is enough inherent nutriment in the moss to give the plant a start. Several times a season, dilute liquid fertilizer may be applied, especially as the plant grows into the second and third years. There seems to the writer to be many advantages to this method of growing plants, and more especially in the fact that transplanting to the garden with its ball of moss insures that the plant will not die during the first season for lack of moisture. Such a plant-growing method merits more general trial, especially for many difficult and acid-loving plants.

In summary the author would certainly rate the sphagnum moss as one of the most useful of our plants, even though such a primitive and humble one.

Chapter XII

LOOK SHARP —— LICHENS AND FUNGI

Lichens

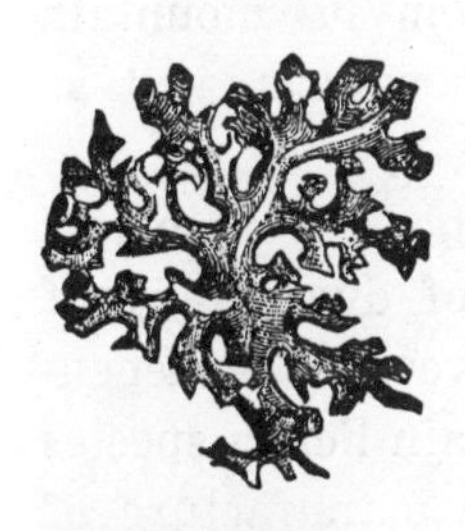

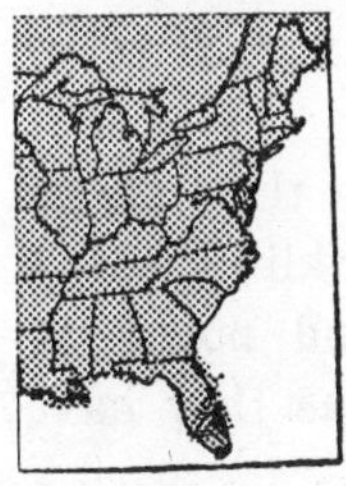

Lichens are one of the most common, most interesting, and yet least observed plants of the world. Whereever there are shady woods, stones, rocks, poor soil, cold weather, or other particularly difficult conditions, one will find the lichen growing. As a form of plant life lichens are especially interesting to study, because without exception all of the many thousands of species have one feature in which they differ from any other form of plant life. Each lichen plant consists of two separate entities, living together in a balanced relationship called symbiosis.

Thus the composite lichen plant is actually a joint relationship of fungi and algae growing together in a physiological union, and the combination of the forms of fungus or algae account for the variability of the many species of lichens. One will find lichens in an amazing variety of shapes, colors, and locations. Some lichens, such as reindeer moss, have common names, but many are identifiable only by a relatively few botanists whose specialty lies in these primitive forms of plant life.

Pertinent to our discussion of useful plants is the fact that in times of famine and in arctic regions, lichens are the source of an emergency food for man, as well as being a regular diet for such animals as the caribou, reindeer, and musk-ox. An exhaustive discussion of these and other uses for lichens is found in the Smithsonian Institute report for 1950. Parenthetically, that monograph offers evidence to indicate that the "manna" of the Bible was a form of lichen which is still eaten by some desert tribes, and which, when blown loose from its mountain habitat by high winds, will roll into the valleys, just as the Bible story relates.

Commercially, other than as a forage, the principal use for lichens seems to be as a source of dye. Many of the soft colors of the favored Harris tweeds are dependent upon the lichens for their color. Certain lichen species are also used at the present time in the manufacture of perfume, while others find a use in the manufacture of soap.

Medicinally, lichens have been used since the time of the Egyptians, even our own Benjamin Franklin making much of the virtue of one form. Dried and powdered lichens are used in cosmetic powders today, as they have been for four hundred years.

For the roadside collector, however, one of the principal uses for lichens might be as a form of decoration, for in combination with other plants on a Christmas table or in a terrarium, lichens are very effective.

Edible Fungi or Mushrooms

Because the word "mushroom" has become firmly attached to one commercially grown genus of fungus, it is perhaps improper to call this chapter a discussion of mushrooms. There are many forms of edible fungi which are just as palatable and often much more desirable than the field mushroom. People try to differentiate between *mushrooms* and *toadstools*, the term "mushroom" being applied to the edible and "toadstool" to the poisonous species. This again is not a scientific distinction, and to be correct we should learn to look at the various forms of the *edible fungi*, each as a separate entity.

It is true here as with almost any form of nature that with the fungi there are good and bad species. And just as the public is generally quite apprehensive of even the most harmless garter snake because there are poisonous snakes, so with a fungus like the mushroom, (which has some resemblance to the deadly *amanita*) many people are extremely wary of using any form of fungi for food. This is surely an unreasoned fear, for it is relatively easy to distinguish between most of the edible and the poisonous forms, though this does not mean that one should not be careful in selecting mushrooms for eating. If one does not know which is which, it would be better to restrict the gathering of fungous growths to something other than a species that just looked like the mushroom, because it is true that the deadly form grows in the typical umbrella-style. However, since it is much better to be safe than sorry, sufficient knowledge for safety should be acquired.

One caution which is worth giving is to avoid the use of any of the so-called "tests for poisonous varieties." The familiar silver spoon method of distinguishing the poisonous from the edible forms is definitely unsafe, as are any of the other "old wives tales." One of the safest ways to know what you are picking is to ask someone who has had previous experience. Full protection from mushroom poisoning will, however, be gained by following these six suggestions:

1. Never eat a mushroom until you are sure of its identity.
2. Be sure all mushrooms are fresh.
3. Do not use brightly colored forms.
4. Reject all forms that have a cup or sack-like envelope at the base of the stem.
5. Reject all forms in the early, or button state, for at that stage their characters are not distinguishable.
6. For anyone who should become enthusiastic about the subject of mushrooms, there is a wide choice of reliable literature which will be found in any public library, and also a number of government bulletins which delineate the values of the various forms.

Agaricus campestris — FIELD MUSHROOM

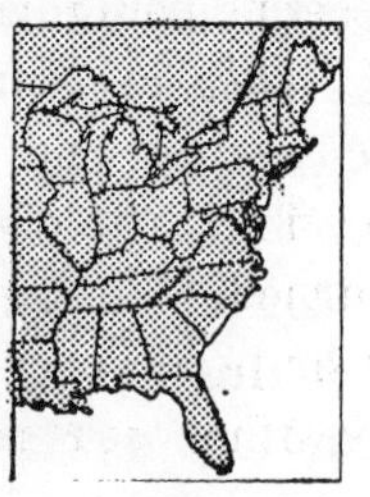

Here is the mushroom which is closely related to the market form. There are several species of *Agaricus* which spring up on lawns after rains during the growing season, but the difference is largely a botanical one. It can be distingushed from the poisonous form which is similar in appearance by the fact that the cap of the edible mushroom should be white or pinkish, with the texture of a kid glove. The stem should be white and solid with a ring or frill slightly above the middle, near the top. The gills underneath do not quite reach the stem, and they should be whitish in the early stage, and pink when the mushroom is at its best. The flesh itself is white, turning pink when broken, and there is a characteristic mushroom scent always present. There would be no bulging at the base of the plant stems, such as is present in the poisonous species.

In addition to the normal methods of cooking the mushroom, the surplus may be dried when large quantities of them are found, which happens sometimes under the proper weather conditions. Any edible fungus to be dried can be peeled and sliced, and the pieces dried over the stove on paper, or strung on thread and dried in a cool oven with the door open. Store in a tight can away from dampness. They should be soaked in tepid water and then used as if fresh. Such dried mushrooms imported from Europe are now available in most city super-markets.

Morchella species — MORELS

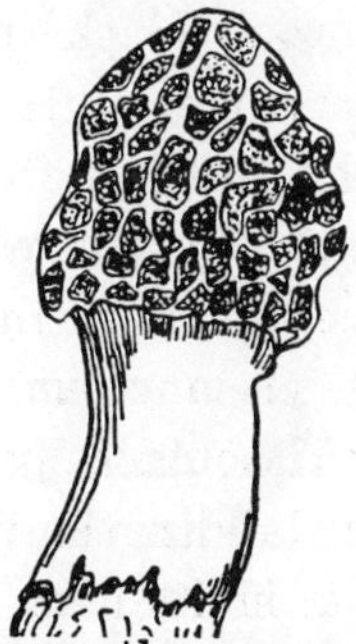

This is one of the more interestingly shaped of the edible fungi, looking when grown quite like a sponge on a stick. The pebbled or granular surface grows on a stout hollow stem, the cap and the base being inseparable. In height they grow from two to six inches. There are a number of species of morels, all of the genus being ranked among the very choicest of edible fungi. They are usually, however, not found growing in groups, and so one must depend a great deal upon good luck for finding a meal of morels. Morels usually appear on burned-over ground, and also appear frequently in old apple or peach orchards. Morels are to be cooked much as any other mushroom, and have a distinctive flavor.

Coprinus micaceus — INK CAP

Sometimes early in the spring before any other wild mushrooms are available, then appearing intermittently throughout the season right up to the time of frost and always after any extended warm period, one can find great masses of the ink cap growing around or over the roots of dead trees.

The ink cap will be found growing in dense masses, each on a stem about an inch and a half high, in shape like a half-opened umbrella. The underneath gills are pale, becoming pinkish and finally black, with the rapid growth from appearance to dissolution into a black inky mass being the process of only 72 hours.

Ink caps should be picked just when the umbrella is half opened, and quantities of them may be cut with a knife or scissors in such a manner as not to include the dirty base of the stems. Wash well in a colander, drain, and cook without adding water. The black juicy mixture is deliciously flavored, and after adding butter and seasoning, may be thickened with a little flour to make a creamy sauce for use either on toast or as an ideal addition to a steak dinner. Two quarts or more of fresh ink caps will provide a good "mess" for the table, but such a quantity is not hard to come by since there are usually plenty in one growth. Watch for ink caps as you drive through the country and gather them as soon as you see them.

Pleurotus ostreatus — OYSTER MUSHROOM

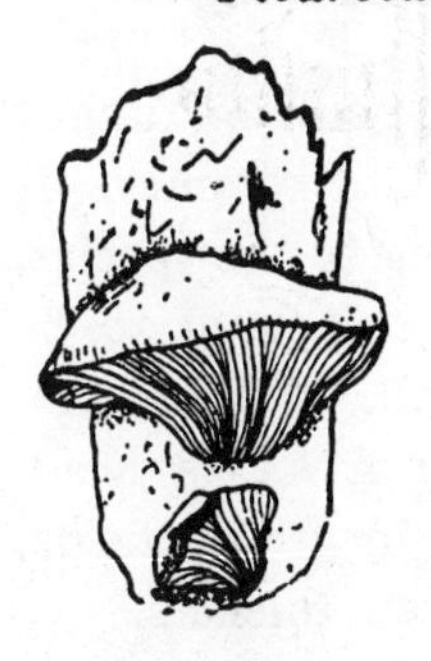

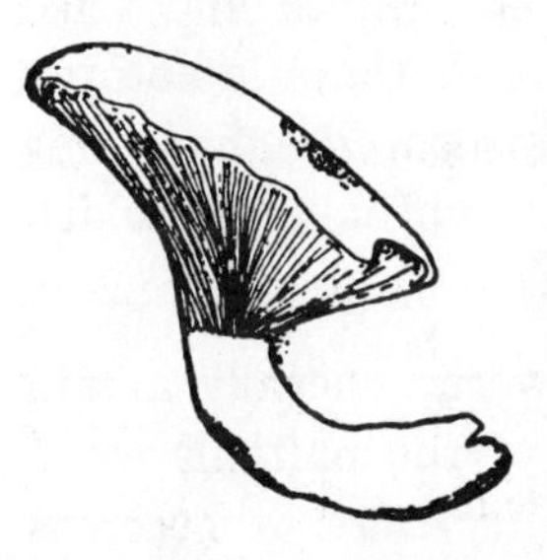

Like the ink caps just discussed, this fungus seems to be most happy growing on an elm tree, although it may at times be found on poplars. Unlike the ink cap, however, it is found not on decaying roots but only in the rather high crotches of a dying branch, and it is usually not easy to secure, except with a ladder or long pole. Anyone living in a section where there are many people of Italian parentage knows how eagerly this growth is gathered by these Latin gourmets.

This particular genus is most easily found from about the middle of September through November, but because it grows to a size of from twelve to fifteen inches in diameter, one specimen is enough for a large meal. If not gathered when real young they are somewhat woody and leathery in texture, but cook up easily. One Italian expert, writing in *Yankee Magazine* recently, offered the following recipe:

> "After washing and rinsing cut into one inch cubes. Place cubes in pan of cold water, just enough to cover mushrooms, let come to a boil, and simmer for one half to an hour until tender. Then drain and dry between towels and place in a pan containing olive oil or other cooking oils. A kernel of garlic added will give more flavor. The pieces should be fried until brown, and will thus provide an unusual meal of excellent flavor."

Lycoperdon gigantea — GIANT PUFFBALL

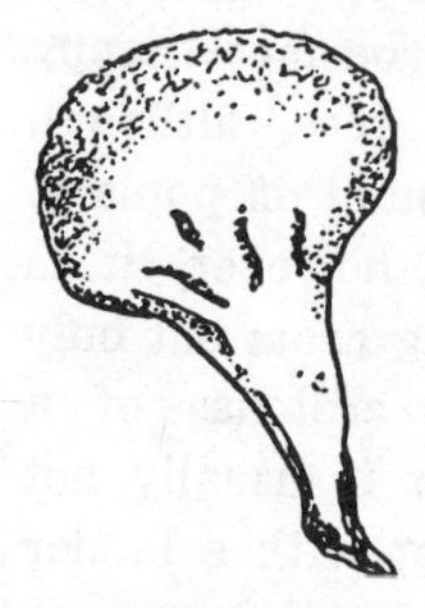

The puffball is an edible fungus which is found only in late summer. It will vary greatly according to the species, from tiny round forms too small to be used to the giant puffball, which sometimes attains the size of a man's head. They are found growing in grassy fields, along roadsides, or in rich thickets.

When found, they can be peeled and sliced and fried in butter, and in the opinion of the writer there is nothing better, nor anything more easily prepared, than this delicacy. One should use only those puffballs in which the flesh is white and firm.

It is said that the Indians, who were generally afraid of all forms of fungi, did make use of the puffball when it matured. Following the edible white stage, it becomes a great brown cannonball, containing billions of spores, and dark brown in color. These spores were known by the Indians to have styptic properties and a supply was kept on hand to be dusted on wounds. Such a medical use is confirmed in the discussion of herbal plants by Fernie, an English authority. He suggests also that puffball spores would make a fine drying powder for dusting on sores between the fingers, the toes, and under armpits.

Gerard, the great herbalist of the seventeenth century, discusses at some length the interesting fact that the dark brown powder inside the puffball produces a flash like lightning when blown at a fire. In Scotland the puffball is called the "blind man's eye," because it is believed that the dust is so fine that it will cause blindness. Perhaps it would.

. . . look like the innocent flower, but be the serpent under it.
Macbeth, Act I, Scene V

Chapter XIII

LOOK OUT —— POISONOUS PLANTS!

In spite of the great many thousands of species of plants in which our territory abounds, there are comparatively few which are poisonous, just as in the fauna in this region there are only a few objectionable animals. A few plants are as dangerous as the rattlesnake and the wildcat, or as much to be avoided as the skunk. But with only a little care and knowledge, they may easily be shunned. Some are so extremely common that knowledge of them is essential and these and a few other plants are noted in these pages. In any case, there is one thing certain about a plant, — unlike an animal, it will not jump up and hit one or bite or scratch one's eyes out, and hence with poisonous plants, *"knowledge is power."*

In any field of thought selective lists are always subject to criticism, and the writer is well aware of the fact that some might criticize his selection of poisonous plants and would suggest that there are others that should be mentioned, but — right or wrong — these seem to be the most important from the viewpoint of the average tourist or roadside traveler. Some, as indicated, are poisonous to the touch, while others are poisonous only when eaten, and several are perhaps poisonous only to the specific person.

Aconitum (species) — MONKSHOOD

An alternate common name for this plant, especially the garden form *A. Napellus,* is "deadly aconite", which has been used for everything from murders to rat extermination. Rare is the well-ordered perennial border in the home garden which does not have one of these beautiful, rich-blue flowered plants. Its deadly properties have been known for centuries, mention of *it* having been made by Ovid, Nicander, Theophrastus and other writers of Rome. It is a border plant which should be placed where children will not accidentally chance upon it, as even a nibble of the leaves could cause serious damage.

Actaea rubra — BANEBERRY

Another name for this plant is the snake-berry, which is a somewhat appropriate name, considering its poisonous nature. The other common name is the necklace-berry, given to it because of the rather beautiful red and ivory-white berries, which are borne in elongated clusters in summer. The leaves are compound, made up of three sharply toothed leaflets. Caution against this

plant is necessary because, the fruit being so beautiful, it is hard to understand that poison could reside in such a pretty package. Admire the berry all you will, but avoid its fruit.

Aesculus Hippocastanum — HORSECHESTNUT

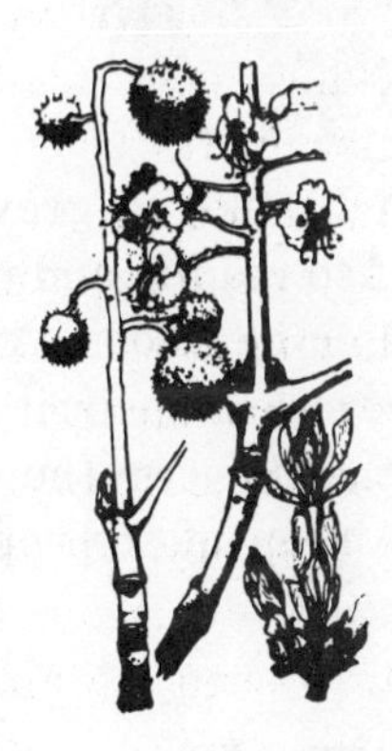

It is quite possible that one of the young child's first contacts with plants is the gathering of the newly fallen, lovely, shiny brown nuts of the horsechestnut — so good to collect and to use as toys and "trading stamps." Their taste is bitter, and it should be noted that, eaten in any quantity, they are quite poisonous. Do not confuse horsechestnut with chestnut *(Castanea)*, an entirely different tree.

Aethusa cynapium — FOOL'S PARSLEY

This plant so greatly imitates the foliage of the carrot and the leaves of parsley that it is hard to indicate variances that would be appreciated by the novice. The rule seems to be, therefore, that one should never pick, to use as parsley, the foliage of any wild plant, because it well might be this plant which contains an active stomach poison.

Ailanthus altissima — TREE OF HEAVEN

This famed "tree which grows in Brooklyn" grows so easily in difficult places, and seeds itself so readily and often, that it is found along many roadsides or in woods. An encounter with its foliage may cause severe skin irritation, but the smell of the foliage and of the flowers (however handsome these may be) is enough to keep most people at arm's length.

Amanita viscosa — DEATH CUP

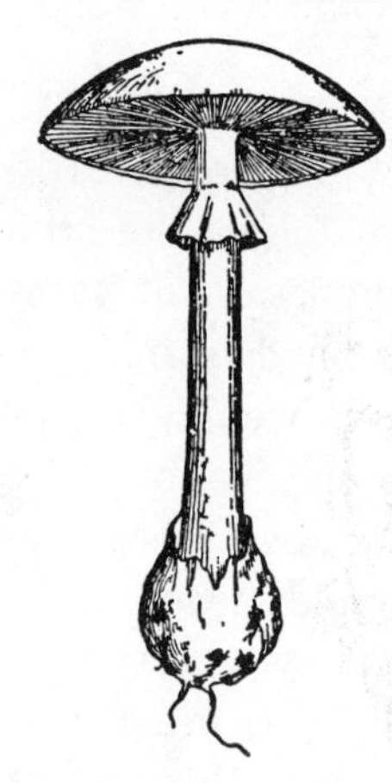

Here we have what is possibly the most poisonous of all American plants, and certainly the most poisonous of the mushroom-like fungi. The plant is dangerous even to handle, but at least there is the advantage that the several poisonous species are fairly easy to identify. The principal identification of *amanita* is found in the large cup or bulb at the base of the stem, and the ring or skirt near the top of the stem. The two most common poisonous forms of amanita are

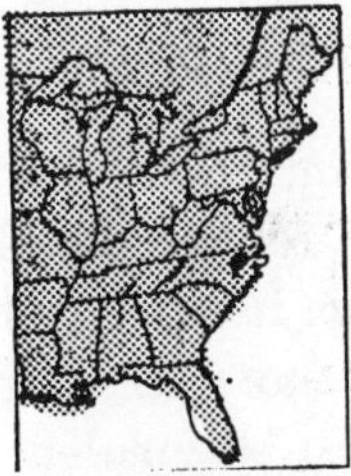

the *death cup,* with the top of the cap varying from pure bright white to a pale yellow, and the *fly-amanita,* with the top of the cap varying from yellow to red, the surface of the cap being flecked with scales or white spots.

As indicated in the discussion of edible mushrooms, one should never eat fungi about which there is any question, nor any of those which are found in the button or unexpanded state. Avoid, too, any mushrooms which have the under side of the cap full of pores, or those containing a white milky juice. Equally to be shunned are all woodland mushrooms with a flat top, or extremely bright colors, as well as the yellowish orange mushrooms which are growing on old stumps, however beautiful they may be. Another rule is to avoid any mushroom which is apparently grown past its best, and as always, try to have someone who knows show you which ones to gather.

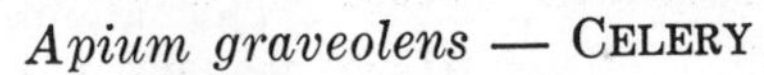

Apium graveolens — CELERY

The purpose of this paragraph is to point out the dangers inherent in such an ordinary garden and table plant as common edible celery. To some people, the foliage of celery is a skin irritant similar to poison ivy, thus indicating that the entire family of Umbellifers contains, to a greater or lesser degree, a poison which, at worst, can be virulent or, at best, merely irritating. It is precisely this poison content that makes the garden parsnip inedible until frost has changed it into a tasty sweet vegetable.

Apocynum cannabium — Dogbane

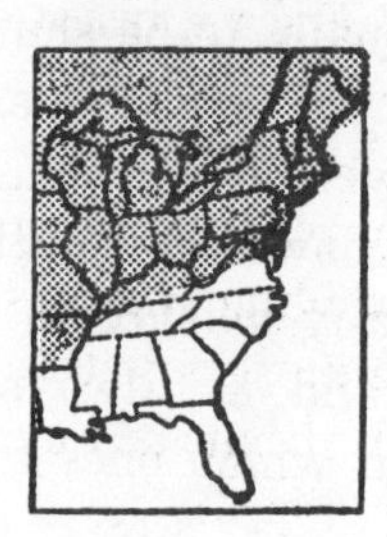

There are several species of the dogbane, all of which, because of the milky juice in the stems and leaves, might be confused with the young growth of the milkweed. The stems, however, are tougher and forked in branching habit, a circumstance not true of the milkweeds. Another name for this plant is the hemp dogbane or Indian hemp, given because of the fibrous stem. The poison in the plant is found in the milky juice, but as this juice must be taken internally to cause trouble, the dogbane is not considered a poison plant of first importance.

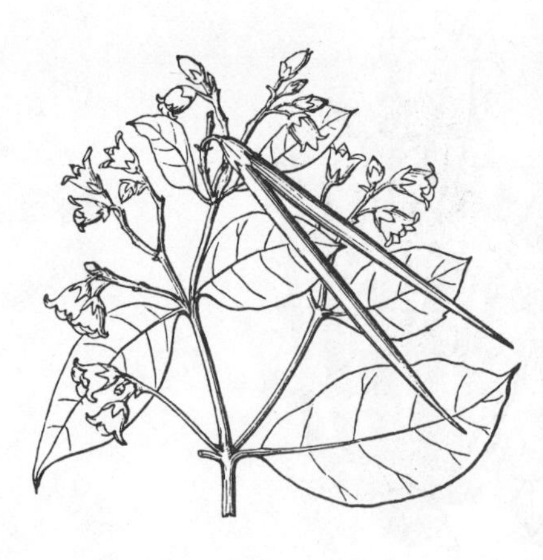

Atropa belladonna — DEADLY NIGHTSHADE

Most readers probably know the name belladonna as something used by the physician, but may not realize that this plant is often found growing in the wild, or in old gardens. It is useful in a low concentration, but in larger doses it produces hallucinations and, eventually, death. The plant bears a fruit that, because it looks like a small black cherry, is apt to attract the child who likes to try anything that looks edible. Consequently, if you find a plant with dull green leaves, rambling purplish stems, and purple flowers with a touch of yellow, pull it out, however attractive it may appear.

Cannabis sativa — MARIHUANA

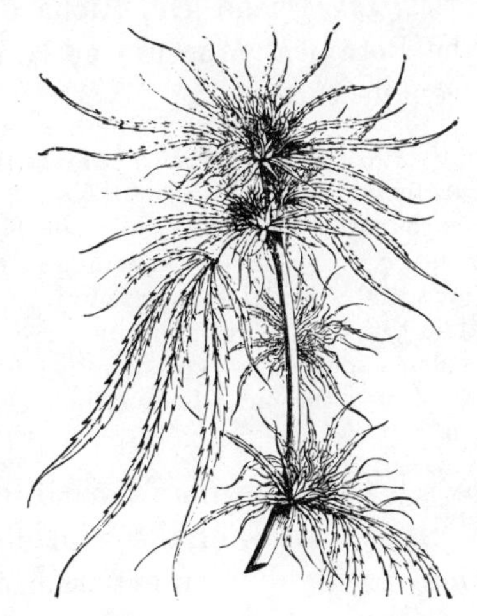

This long-used but recently much-discussed plant is quite often found growing as an escape in fields or, more probably, in a city dump. The other common name of the plant, Indian hemp, indicates its country of origin. There, as in other parts of the East, it has been known for centuries under the name of hashish or *bhang*. The persistence of its use suggests that it might have had a partly ceremonial purpose, as did tobacco (now known to be poisonous) for the American Indians. Whereas the American Indian used tobacco as a ceremonial, medicinal, and relaxant plant, we have now come to know it as a deadly drug, and to realize the truth of the old saying, "Tobacco is an evil weed — and from the Devil did succeed."

In its uninformed exuberance, the younger generation has abused marihuana, referring to it as "pot," "tea," "weed," or using it in "reefers," and suppression of its use has become a major headache to our police. Its advocates claim that it is not dangerous, or that it is an adjunct to religion of a sort. What should be the approach of the average citizen?

I reply to this question by quoting a recognized authority on narcotics, my friend Dr. Richard Evans Schultes, Curator of the Botanical Museum at Harvard University, who, in a recent lecture, said:

> We should always look at narcotics against the social background in which they are employed. One example will, I think, show what I mean. In Moslem countries where the use of alcohol is highly unacceptable, *Cannabis indica* or hashish — marihuana to us — is taken by millions with no social stigma. In our country, where the generally uninformed public throws up its hands in horror at the use of marihuana, millions drink alcohol daily, many in too great an excess for our mechanized society; the habit is usually accepted as a highly respectable facet of our culture.

Thus we see that one man's meat may be another's poison. Some of the plants of the wild discussed so far in this chapter may be highly virulent when ingested unknowingly and in quantity, but if taken under supervision may be medicinal. Or, as is the case with marihuana, some plants may be the source, in a given culture, of deeply affecting experiences or, taken in another cultural context, of disturbing, hallucinogenic ones.

Cephalanthus occidentalis — BUTTON BUSH

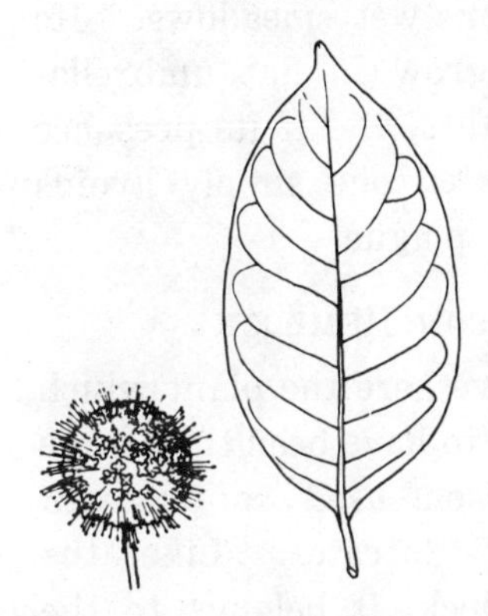

This large shrub, usually found growing in wet places, bears interesting button-like seed cases that are reputed to be quite poisonous if chewed and swallowed.

A plant such as this illustrates the importance of studying botany, and of familiarizing oneself with the habits and properties of not only the plants of the wild, but with those of the more immediate home landscape as well.

Cicuta maculata — WATER HEMLOCK

This is one of the most violently poisonous of our wild plants if taken internally. Here again, as with *Aethusa,* it has leaves which resemble those of the parsnip or of overgrown parsley, while the thick roots also have a parnip-like smell. Sometimes the roots are brought to light by the action of frost which heaves them out of the ground, and therefore children who play in the woods should be cautioned about eating anything which has such a root.

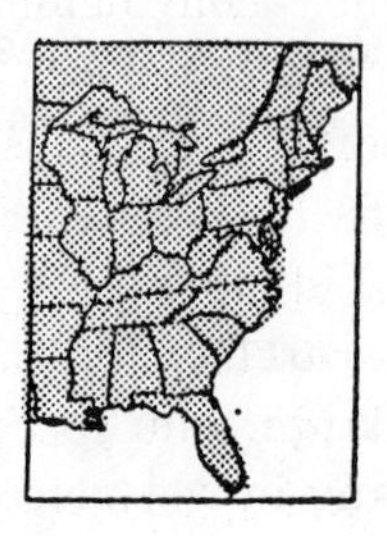

Another name of the plant is beaver-poison, which shows that it possibly has poisoned animals as well as man. It is never found growing except near shallow water, in swampy areas, or wet meadows. Its flowers grow in flat umbrella-like clusters, and if its presence is suspected one simply avoids it like a plague.

Conium maculatum — POISON HEMLOCK

We have here the plant which is supposed to have been the source of the poison used as a death potion for Socrates. Like the water hemlock, it belongs to the parsley family. The foliage is very fern-like and, if one can be objective, really beautiful. In appearance, the plant suggests a large tree-like carrot, and bears large flat umbels of white flowers.

Unlike the water hemlock, it is more likely to be found growing around waste lands, old dumping grounds and roadsides, than along the edges of streams. It is found variably throughout the eastern and central states and more generally around the coastline ports. Unlike the roots of the water hemlock which produce

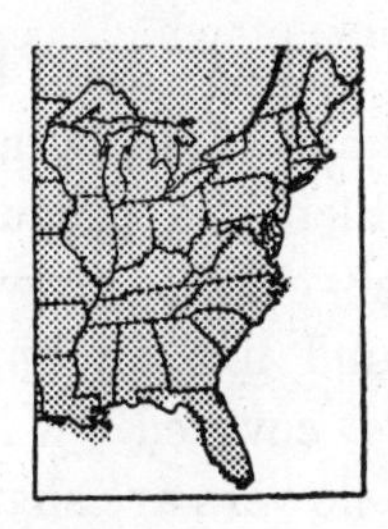

only violent convulsions, the poison hemlock produces an almost immediate paralysis which results in death. We can say only what was said about its relative — that it is better not to try to gather any plants or seeds which resemble parsley or carrots or any seeds supposed by the novice to be caraway or anise, since there is a great possibility that they may be "poison hemlock."

Daphne Mezereum — DAPHNE

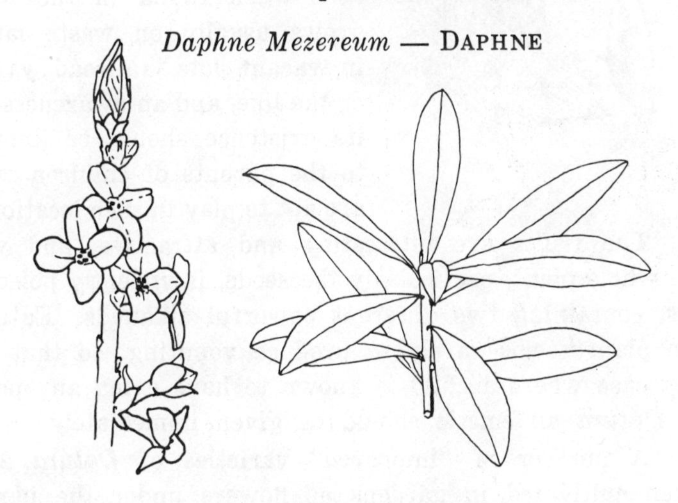

Almost the first among the flowering shrubs to bloom is the fragrant, lilac-colored daphne. Although there is small chance that anyone will want to nibble at it, it should be known as poisonous. Other members of this family, such as the very lovely *D. Cneorum*, are harmless.

Datura Stramonium — JIMSON-WEED

Another name for this plant is the thorn apple, because the fruits which are produced on the rather coarse growing plant are round like an apple, and heavily covered with prickles. The flowers are shaped somewhat like petunias, with the color ranging from white through shades of violet. *Datura* when found in the wild grows usually on waste land, in vacant lots, railroad yards or the like, and an awareness of its existence should be known to the parents of children who are apt to play in such locations.

The fruits are interesting and attractive, and yet the whole plant, particularly the seeds, is violently poisonous, containing two different powerful alkaloids. Eating the plant seems rarely to produce vomiting, so that in any case where a child is known to have eaten any part of *Datura* an emetic should be given immediately.

A number of "improved" varieties of *Datura* are often cultivated in gardens as flowers under the name of Angels-trumpet, and therefore gardeners should be aware of the poisonous nature of elements present in this otherwise pretty plant.

Dieffenbachia Seguine — DUMB CANE

This is one of the "good" and easy house plants — good because of its easy growth, fine foliage and tolerance of heat and dryness. But the common name of dumb cane has not been assigned without reason by the natives of South America, as the exudation from the stem, if taken internally, paralyzes the power of speech. Buy and use these fine plants in your home, but don't serve them as food, not even to a mother-in-law.

Euphorbia pulcherrima — POINSETTIA

It has been both stated and refuted that the Christmas poinsettia is poisonous, but as the chance of anyone's sucking any of the milky sap in the home is very remote, these plants should hardly be excluded from the season's decorations. Yet it is mentioned here because all of the Euphorbias or Spurges have the same milky sap in their stems, and a number of species *are* poisonous. The rule is to beware of any plant at all that exudes such sap — insofar as ingesting the fluid is concerned!

Holcus lanatus — VELVET GRASS

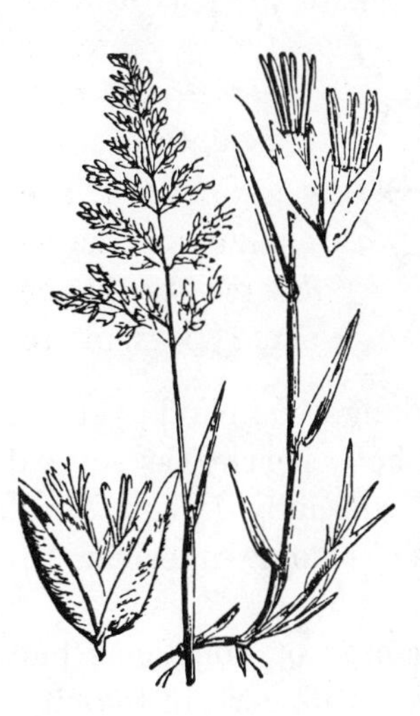

It is rather difficult to realize that not all forms of grass are edible, because we are in the habit of seeing animals pasturing widely on grass of any kind. There is one grass, however, which contains in small amounts the virulent poison, hydrocyanic acid. This is the velvet grass, which, because of its beauty, often tempts one to pick it and chew the stems. As most everyone is subject to this impulse, the appearance of the grass as shown in the illustration should be noted.

Hypericum perforatum — ST. JOHNSWORT

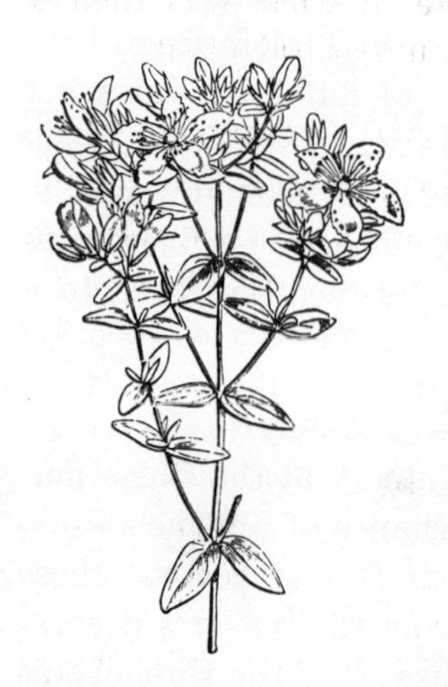

There are a number of the common wild plants (weeds, one might call them) which, while not actively and immediately poisonous, still should not be eaten. A good example is the widely found St. Johnswort, where the active principle found in the plant's chlorophyll does not break down in the system and may have serious effects in the course of passing through the liver.

Similar effects may be caused by eating the foliage of common cultivated buckwheat, but this is not likely to happen. The dangers inherent in plants such as these are similar to the dangers of crossing the street and being run down; quite possible but not probable, given reasonable caution.

Ilex (species) — HOLLY

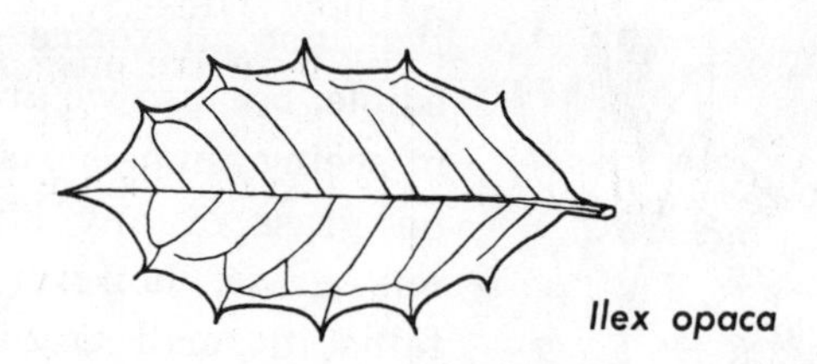

Rarely is it said that holly is a dangerous plant; yet one would not want to eat a quantity of holly berries, as several members of the American hollies, especially *I. vomitoria*, were the source of the "black drinks" of our American Indians.

Mistletoe, the Christmas companion of holly, has had a bad reputation, and indeed it does have a certain toxicity.

But its exclusion from churches in medieval times was probably because, growing as a parasite in what was then a mysterious way, it seemed to be an unnatural plant.

Ipomoea (species) — MORNING-GLORIES

In the course of youth's experiments with and search for new forms of hallucinogens, someone uncovered the fact that, in South America and elsewhere, the seeds of some of the various species of morning-glories have been used to induce pleasurable or exciting states of being. A leading authority on hallucinogens tells me that this property is found not in seeds of the commonly planted *Ipomoeas*, but only in tropical varieties which are not hardy in our climate.

Iris versicolor — BLUE FLAG, POISON FLAG

"Beauty is only skin deep" could here be paraphrased as "Beauty is only on top," for in the case of this very lovely iris of the verges of ponds and streams, the roots (rhizomes) are definitely poisonous, as indicated by one of its common names. The acrid resins found here are purgative and diuretic. Note that the underground parts of *any* iris are suspect.

Lantana Camara — LANTANA

With increasing frequency one sees the brilliant and perpetually-flowering plant we call lantana in gardens in the North, although we do not find it appreciated in the South, or in Hawaii, where it is a bad weed. One finds records of deaths caused by the eating of the foliage, although the generally

spiny nature of the plant would not encourage one to partake. It might be well to alert little children to the bad character of the plant, if you grow any.

Nerine oleander — OLEANDER

Not all of the poisonous plants are to be found along the waysides. Among the house plants commonly planted in the subtropical areas of our country, and especially desirable in tubs for the terrace is the lovely flowered oleander, yet there are numerous reports that the chewing of the leaves of this plant have produced fatal results. Especially to be avoided is using the oleander's hard-wooded shoots as meat-skewers for cookouts — something so easy to do in the case of a plant already growing on the patio.

Philodendron (species) — and other ARUMS

To the great and lovely family called Araceae belong not only the Dieffenbachia discussed earlier but also a great number of plants undesirable or unpleasant — the skunk cabbage and the jack-in-the-pulpit, for example. All of this family, including a great number of species of the philodendrons, should be viewed with respect. In varying degrees, they all contain what are called calcium oxylates which, taken in the mouth, produce disastrous effects on the throat.

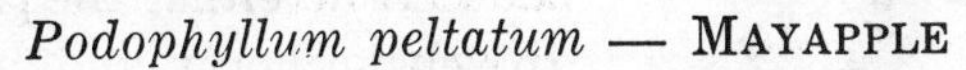

Podophyllum peltatum — MAYAPPLE

The mayapple, that lovely foliaged plant of the deep woods, is a medicinal plant whose uses were known to the American Indians and transmitted to the settlers. It first took the name of 'wild jalap' as it is a purgative plant, and is a medicinal constituent of cascara compounds. So while it is not violently poisonous, it's not a plant to be ingested thoughtlessly. Of its various parts, the fruit is the least harmful and has been used in the juice of punches, or for jams.

Primula obconica — PRIMROSE

It is not "the primrose by the river's bank" which is being noted here, but rather one of the nicest of the varieties grown by the florist, *Primula obconica.* This bloom contains a contact poison that can be very virulent to some people. But here, as with poison ivy, knowledge is power, and if one is "respectful" there is no danger — with primroses or with any other plant.

Prunus (species) — WILD CHERRIES

On page 126 I mentioned the tasty, beautiful jelly which can be made from either the rumcherry or the chokecherry and, also, that an extract from the bark was used as an ingredient in cough syrups. Both statements are true, and yet we have to add a skull-and-crossbones to these names, as the high concentration of prussic acid or cyanide found in all parts of the trees places wild cherries among the poisonous plants. Cattle have been known to die after browsing these trees. But this does not mean that one should not use the fruit as suggested. The truth is that a similar kind of poison is found in the seeds of apples and the pits of peaches, but these are parts not usually eaten.

Rheum Rhaponticum — RHUBARB

In the garden or in an abandoned country farmyard one finds the very delightful and edible rhubarb, a plant one rarely thinks of as poisonous. And yet one should never, under any circumstances, think of cooking it as one does spinach — that is, using the developed leaves, for these are definitely dangerous to eat. Rhubarb stalks, yes, but go no further.

Rhus Toxicodendron — POISON IVY

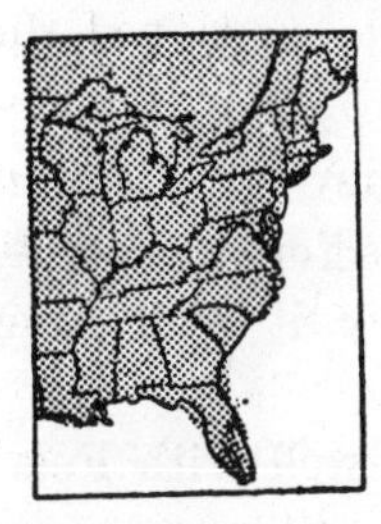

So many warnings have been issued and so many people have accidentally become poisoned that it is hardly necessary to mention this most common of American poisonous plants. There should be but little trouble about the identification of poison ivy, because of the three distinct leaflets on a single stem, which differentiates this ivy from other forms. Another characteristic of the poison ivy is its beautiful glossy, dark green leaves, as well as the beautiful shades of red which these same leaves attain in late summer and fall, which in their brillance provide some of the finest colors of our fall landscape. One indeed could well recommend the poison ivy as a landscape plant, if it were not for the extremely dangerous oil, which the plant contains.

In some sections of the country one finds the Poison Oak, which is but a variant of Poison Ivy, having the same tri-foliate leaf growth. The difference is found in the rather more shrubby growth of the poison oak and the softer oak-leaf-like appearance of the foliage.

There are some people who seem to be immune to Ivy-poisoning, but it is never safe to count on any such immunity, as it may expire at any moment. At times, persons who for many years have shown immunity will, for instance, be poisoned by smoke from a brush fire in which poison ivy is burning. It is said that even the pollen of the flowers is dangerous to some people and in every way poison ivy is a plant to leave strictly alone unless one is extremely well protected by gloves and clothing.

As to the cures for poison ivy — they are legion, and yet none seems to be entirely efficacious for all people. When one suspects exposure to poison ivy it may save misery to follow these suggestions:

1. Wash the exposed parts extremely well with Fels Naptha or any similar soap.
2. Wash all exposed parts of the body with gasoline or benzine. The oil-solvent action of gasoline helps dissolve the dangerous oil which causes the poisoning.

If it is thought that the exposure has been severe, it is recommended that a doctor be consulted at once, for ivy poisoning can be extremely dangerous and painful, as well as long drawn out in its effects. Never look on ivy poisoning lightly.

Rhus Vernix — Poison Sumac

This is a close relative of poison ivy, which will be found growing as an attractive shrub in localities where also will be found its close relative, the more beautiful and useful Red sumac. While this shrub is not as virulent as poison ivy, the poison sumac should be avoided. The picture will give an idea of the appearance of the green leaves and the almost pure-white berries. Identification can be made by noting the compound leaves, and the fact

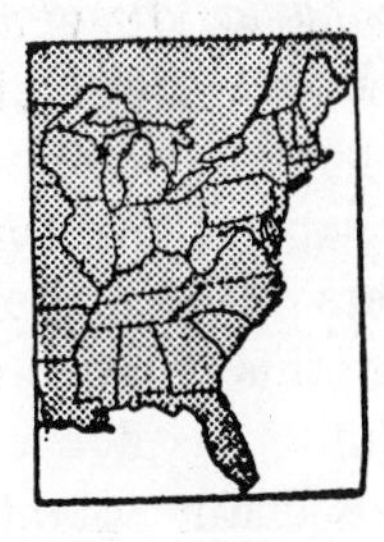

that it seems to prefer growing in a rather wet, boggy, and acid soil. The berries are in themselves quite attractive, but they are highly toxic and for this reason any contact with the plant should be avoided.

Ricinus communis — Castor Bean

If children who have been doctored with castor oil were told that it came from castor bean plant, they surely would have nothing to do with it. But the extracted oil is more medicinal than poisonous, and the plant would not be mentioned here except for a caution about the lovely, varicolored beans from which the oil comes. For the truth is that if these attractive beans are eaten as is, they are highly dangerous.

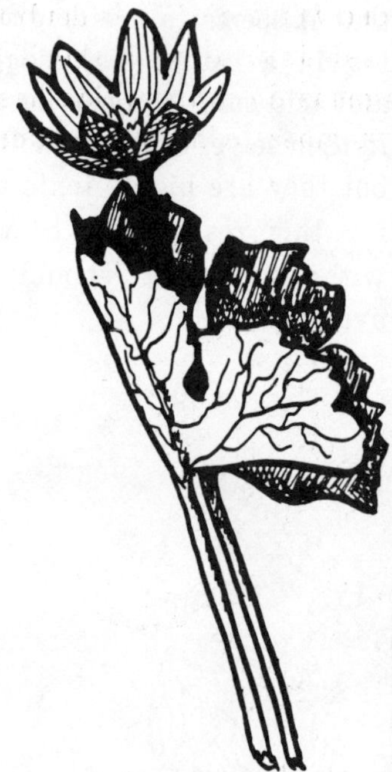

Sanguinaria canadensis
BLOOD ROOT

Any person who has been out in the early spring to gather wild flowers needs little introduction to this very beautiful native plant. It is not a deadly poison like many plants, but anyone who should attempt to eat it in any way would find that its rather succulent looking root was not only bitter, but if taken in quantity would cause vomiting and possibly later paralysis. Gathering and enjoying the flowers does no harm. In proper quantities and under medical control, the poisonous principle has been put to good use as a stimulant and emetic.

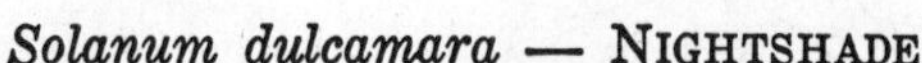

Solanum dulcamara — NIGHTSHADE

The berries of this plant when they finally turn a ruby-like red are most attractive to children but they are not for eating under any circumstance. Solanum, a relative of the tomato, grows in this species almost like a shrub with dark green leaves and lavender to purple flowers. A nice plant to admire but stop there.

Solanum Pseudo-Capsicum — CHRISTMAS CHERRIES

As a former florist, the writer well recalls that customers would often refuse to buy the bright red Christmas cherries at the holidays, because of the legend that "if they were taken into the home someone would die." It is doubtful that this superstition has any basis in fact, yet it should be noted that the plant is of the genus *Solanum* which has a number of disreputable family members, some of them mentioned on other pages.

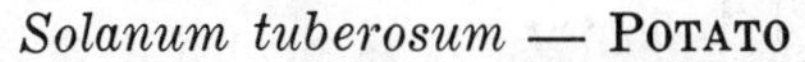

Solanum tuberosum — POTATO

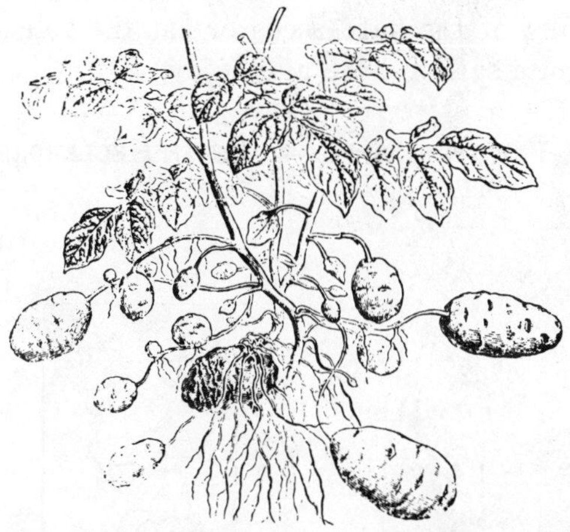

Surely, the reader must say, the common potato can't be a poisonous plant! But it does become one if allowed to stand in the sun for some days after having been dug, at which time the skin will appear green. Such a green potato, as well as any green shoots which may arise, is rated as a deadly poison, not to be eaten cooked or green.

In this connection it is well to remember that this is a member of the *Solanum* family (p. 224) which includes not only tomatoes, peppers, and eggplants, but also such things as angel's trumpets, deadly nightshade, henbane and other really dangerous plants.

Taxus (species) — THE YEWS

Knowledge of the poisonous nature of the yew berries, so attractively red in the late fall, is quite common, but on no account should this fine evergreen fall into disrepute because of this story. The truth of the matter is that the actual tiny seeds *are* poisonous, as is the foliage itself when browsed by cattle; but the soft pulpy seed coating seems to be harmless as well as tasteless. It is doubtful that children will chew many of the hard seeds or eat the foliage so a yew tree is quite safe to have and enjoy.

Veratrum viride — White Hellebore

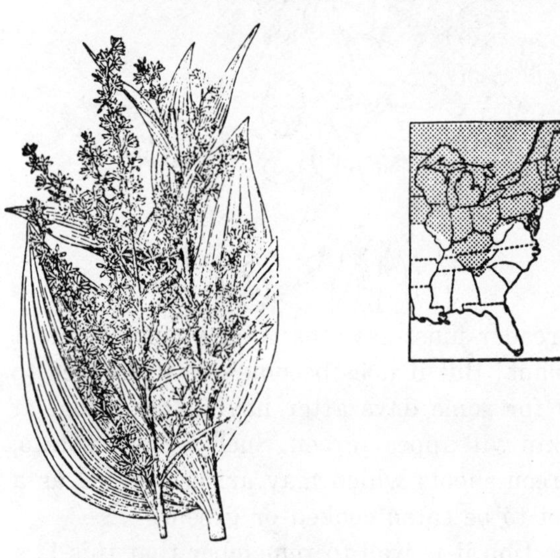

This plant would be but a minor menace if it only grew by itself, for its habitat is solely in inaccessible swamps. It must be mentioned here because it is often found in these swamps growing in association with such useful plants as the edible marsh marigold, and the skunk cabbage. It can easily be identified, even though somewhat similar in appearance to skunk cabbage, by the vertical veining of the leaves which additionally are somewhat hairy. If one were out gathering any of the other plants mentioned, care should be taken not to include even the smallest piece of hellebore, because the leaves contain a violently poisonous alkaloid, which has been known to cause death, while smaller quantities cause heart and respiratory reactions. Hellebore does, however, cause vomiting, which means that it may have its own cure in itself. Remember that *Veratrum* is rated as one of the more poisonous plants of our section, and that it should be carefully avoided.

Wisteria (species) — WISTERIA

Named after a Philadelphian, wisteria as a vine came in more than a century ago, and hence one often finds old plants of it on abandoned properties or in our gardens. However lovely it is, and however tactually and visually attractive the seed pods may be, children should be warned against chewing the seeds, as these are very dangerous.

Chapter XIV
WAYSIDE PLANTS IN YOUR BACKYARD

One of the best ways of fully appreciating the plants of the waysides and the woods is to bring them into your backyard. If you happen to live in the country, this will probably be unnecessary, but for urban and suburban families with children, a wild garden constructed in a backyard could prove to be both a happy and instructive place. Even an otherwise "formalized" landscape could contain a little wildflower garden or corner where plants can be enjoyed and studied.

Another way of making such a garden would be to cooperate with neighbors who live either to the side or rear, jointly setting aside a casual spot where various plants of interest could be placed, creating a background of trees, shrubs and flowers for both participants. In a way, this could be considered a kind of rock garden, but one with a minimum of rocks, in which unusual plants collected on weekend expeditions would recall scenes of mountaintops and valleys one has visited.

In arranging for such a garden one must naturally be as informal as possible, planting trees, even if they have come from a nursery, in the sort of design that one would find in the wild. A few birches giving light shade and some evergreen rhododendrons for winter enjoyment would be a beginning, and from there on the garden could proceed in a number of directions. One could become interested in medicinal plants, and make a garden of healing herbs; one could concentrate on fragrant shrubs, fragrant wild flowers, adding some sweet cultivated plants; or given a sunny open spot, one might plan a meadow garden, or a garden of acid soil plants, or even a "specialty garden" such as a rambling assortment of heathers and other members of the great family which we call Ericas, including azaleas, andromedas, arbutus, laurel and blueberries.

Civic minded gardeners could cooperate in community beautification and education by promoting and helping with a nature section in a public park or, as is now being done so much, by arranging for a "nature garden" on the grounds of a local elementary school. Parent-teacher organizations, Garden Clubs, Boy and Girl Scout, and similar youth groups would all welcome such an effort. Plants that children should know about would be used extensively by alert teachers to illustrate lessons in such fields as general biology, nature study, and even in art. Urban parents raised in the country know all too well how little today's city child knows about nature. In making this suggestion for work with school children, I stress the fact that work should be focused on the 7- to 14-year age group; adolescent high school-age pupils have small interest in such matters.

Also, your own interest could be expanded to benefit the community in the areas of conservation organizations, community beautification, and museum-adjunct schemes. Natural history museums badly need some place for showing a bit of "nature in the raw" — something which cannot quite come off in the artificial settings so commonly seen, however excellent these may be. A garden such as this would promote a knowledge of local plant material, its values, its rarity, its beauty.

An entirely different reason for planning and working in a nature garden would be simply because you love plants. The vogue for rock gardens today is not what it was at one time, but by careful selection one could develop a remarkable alpine garden using only native American plants. Our American flora encompass a wide selection of very beautiful plants, the growing of which would challenge the most expert gardener. It's a curious fact that a great number of our "imported rare plants" are actually natives which in the 1700's went to England and Europe where they were developed and improved, and were then sent back to us as wonderful plants from a foreign land. In point of fact, they may be found growing in the woods not too far from your own home. The growing of a simple little wild garden could be a great education to any gardener.

Finally, a legitimate motive for making a natural wild garden in your backyard might be a very personal one — such as the wish for a little place of retreat in the shade, or for a "conversation piece" which would be different from anything in the neighborhood. A garden of whatever kind is, and indeed should be, a very personal thing, reflecting your taste and personality. And from a practical standpoint, such a wild garden might well be easier to maintain than borders, beds and lawns; moreover, if it weren't kept up as it should be, such neglect wouldn't be too conspicuous.

If for any one of the above reasons, the reader would like to make such a garden of wildings, here are some considerations:

1. The matter of soil will demand your careful attention, not only as to possible fertility but as to its relative acidity. Many of the plants discussed in this book are found in or near the woods, and are happiest in soil which is slightly acid, which ordinary garden soil often is not. True, there are certain wayside plants which grow in sweet or limestone soils, but these are not too common.

To discover the condition of your soil, send a sample,

either naturally or slowly kitchen-dried, to your nearest Agricultural Experiment Station, where they will be glad to make a test and report as to what is required to correct the acid percentage, provided that you tell them what you want to do.

2. What kind of specialized nature area are you hoping to duplicate? Is it to be a lightly shaded spot, an open-meadow planting, or that of a stream edge? Or perhaps your home is located in a mountainous section, where American alpine plants would be happy. Types of terrain are many, as are what the scientists call "micro-climates" — those areas which, while only a few feet apart, may offer entirely different conditions. A "low spot" in a garden may have cold temperatures 10 degrees below a spot only six feet away but four feet higher. There is a variance between the conditions on the sunny, as contrasted with the shady side of a huge boulder. All this sort of thing must be considered.

3. In planning and constructing a wild garden one must take careful note of the natural growth in any nearby wild area. Every single and distinct area of our country has its "typical flora" of trees, of native shrubs in the woods, and of the wild-flower undergrowth, which is the ultimate consideration in planting. Oaks are common trees in all parts of the country but the *kind* of oak might vary greatly, as would also the undergrowth shrubs.

In this latter group, some of the specific kinds, many of which would come under our book's heading of USEFUL, are such plants as hazelnuts, choke- or rum-cherries, sweet ferns, sumacs, roses of one or another species, elderberries and blueberries. The shadbushes are tall growing plants which allow light to penetrate to the floor of an area, as do the alders. The choice of such background material will warrant some study.

Where to start collecting the actual flowering plants of such a garden will depend on one's viewpoint. Is the garden to be a truly "native" garden or is it to be of "exotics"?

Does one want wild plants in order to study them, or to try to duplicate a bit of nature, or both?

A good start would be to begin with the comparatively "easy" plants such as the columbines and anemones; some wild harebells and goldthread; and, if you can buy them rather than "snitch" them from the wild (they are on most conservation lists), certainly you will want a group of the lovely wild orchids. There are a number of forms of native American wild iris, as well as a group of fine summer-flowering lilies. One should certainly plan to have flowers for all seasons, beginning with native Virginia bluebells in spring and forget-me-nots, then to bee-balm and cardinal flowers in summer, with a grand show of our so-lovely hardy asters and goldenrod in the fall. This sort of wild garden would be an "American garden" and lovely at all seasons.

A more advanced or international approach might use, in a similar wild garden, many of the shade-loving and beautiful wild flowers of the European Alps, or a collection of the flowers of the Himalayas or of Japan. There are many paths to follow in gardening, and all lead to a real "happiness hobby."

After you have decided on the purposes, the type, and the location of your garden, you will then consider which plants you want to have and, conversely, which ones will be happy in your spot.

Parenthetically, a word should be said on this subject of "happy" plants, for it is a word often used by horticulturists. I have a friend who is developing a "home arboretum" (instead of a wild-garden). She has many rare plants, and she often tells of moving them around from one spot to another in order to ascertain where the plant will be "happiest." To the plantsman, "plants are people," and deserve the same kind of consideration as to home, food, water, and sunlight as does any citizen. Think of *your* plants as people and they will grow the better for you.

In making the decision as to which plants you will want

to have, you should be guided by far more extensive advice than this chapter can offer. Of course, if you obtain permission from the owners, you can go out into some nearby countryside, dig up random plants which suit your fancy, and then plant them along the paths of your prepared wild-garden site. But this may not result in success, as many plants resent such handling. Some wildings just will not grow if transplanted — they must grow "in situ" from seed, or spend a season under controlled conditions. Many plants, also, are under the protection of wild flower societies, or even of state conservation laws which prohibit their being transplanted.

For further information, therefore, the reader is directed to the study of books on the subject, some reliable ones being:

Handbook of Wild Flower Cultivation, Taylor and Hamblin, The Macmillan Co., 1966.

New Way of the Wilderness, Calvin Rutstrum, The Macmillan Co., 1966.

The Wild Gardener in the Wild Landscape, W. G. Kenfield, Hafner, 1972.

In addition, literature can be supplied by the Audubon Society, by the various wild flower societies, and by conservation groups. Also, there are two other books of my own in this series, *Using Plants for Healing,* and *Fragrance and Fragrant Plants.*

The eager wild-gardener may be somewhat discouraged at the outset by the difficulties posed by the "orneriness" of plants not wanting to be moved, or by the legal or physical difficulties involved in moving them from the woods to one's home garden; he may prefer to assemble the collection desired by purchasing it over a period of time. A list of sources of wild plants has very recently been prepared by the Bailey Hortorium of Cornell University, providing names of nurseries which, from coast to coast, offer native plants suitable for growing in their own or adjacent areas. A selection from that list is presented below; write to these people if interested; an asterisk indicates that there is a charge for the catalogue; they'll send you the price upon request.

*Clyde Robin Seed Company, Inc., Box 2855, Castro Valley, CA 94546

*Arthur Eames Allgrove, 281 Woburn Street, North Wilmington, MA 01877

Dutch Mt. Nursery, 7984 N. 48th Street, R-1, Augusta MI 49012

Ferndale Nursery and Greenhouse, P. O. Box 218, Askov, MN 55704

*Orchid Gardens, 6700 Splithand Rd., Grand Rapids, MN 55744

Floyd and Ruth Tate, Rt. 1, Box 61, Seligman, MO 65745

Francis M. Sinclair, RFD 2, Newfields Rd., Exeter, NH 03833

Mincemoyer Nursery, County Line Road, Jackson, NJ 08527

Gardens of the Blue Ridge, P.O. Box 10, Pineola, NC 28662

Three Laurels, Rt. 3, Box 31, Marshall, NC 28753

*Sunnybrook Farms Nursery, 9448 Mayfield Rd. Chesterland, OH 44026

Siskiyou Rare Plants Nursery, 2825 Cummings Rd., Medford, OR 97501

Vick's Wildgardens, Box 115, Gladwyne, PA 10935

Chas. H. Mueller, River Rd., New Hope, PA 18938

Savage Wildflower Gardens, P.O. Box 163, McMinnville, TN 37110

*Putney Nursery, Inc., Putney, VT 05346

*Woodland Acres Nursery, Rt. 2, Crivitz, WI 54114

Wild Life Nurseries, P.O. Box 2724, Oshkosh, WI 54901

In Canada

*Alpenglow Gardens, 13328 King George Highway, North Surrey, B.C.

C. A. Cruickshank, Ltd., 1015 Mt. Pleasant Rd., Toronto, Ont. M4P2M1

It must be emphasized that this short review of the subject of coming to know wild plants through growing them yourself *is* only a "review." The reader will find his greatest enjoyment and satisfaction if he himself engages in a walk down the various paths along which a study of the wild will lead him—such as an absorbing and rewarding interest in botany and plant lore.

Chapter XV
THE NATURALIST CAMPER

The summer weekend exodus from the city to the country is truly a phenomenon of the present decades in our culture; morover, as I have reason to know, it is nothing peculiar to our United States. Throughout the world today, where more and more people are living vertically in apartments, there is a growing desire to explore the horizontal universe of unspoiled nature that may be found within an easy journey.

Thus it is that the habit of camping is a growing interest, or hobby, or relaxation, or restoration — call it what you will. I have seen camping in Europe, in England, and on the island where I live. Here, on Martha's Vineyard, is the site of the country's most-used youth hostel, annually caring for nearly 5,000 young people who, for perhaps the first time, are hiking, cycling, and exploring nature away from home. Increasing, too, are camping sites for four-wheeled families here and everywhere on nearby Cape Cod, and indeed throughout the nation.

The question one can ask about all of this exploration of nature-in-the-raw is: What does one bring home from such trips? Fun? Yes. Fish? Maybe. And a feeling of exploration. But my advice is to decide on just one purpose (it may be very tiny) of any trip, and let other things develop from that one objective. Think of any single aspect of nature that you might like to explore, pursue it, and you will add much else to your enjoyment and education. Perhaps your first objective will be hunting for wild orchids, or photographing spiderwebs, or collecting colored stones. Hunting for these will lead you into many horizon-widening side-alleys.

For the mundane details of how and where to camp one should consult the various organizations that promote not only the sale of equipment but the use of public and private sites. If you are at all interested in the idea of a camping vacation, write for further information to the National Campers and Hikers Association, Box 451, Orange, New Jersey, mentioning this book when you write. Teen-agers should write for information about that organization to American Youth Hostels, 20 West 17th St., New York, N.Y. 10017.

You may also obtain valuable planning material through reading various government publications available from the Superintendent of Documents, Washington, D.C. 20402. Their *Camping in the National Park System* and *National Forest Vacations* are good buys. And from the same source one can obtain a volume that could be termed a "guide to outdoor recreation." This is the Department of Agriculture's 400-page yearbook for 1967, *Outdoors USA*, which is a useful bargain.

And now you are ready for the camping trip. Let me add to the advice you will receive in the books listed above: your camping list should certainly include plastic bags large enough to hold botanical specimens either for pressing or as part of your plant collection, as well as some special plant you may wish to transfer to your home terrarium or to a wild-garden spot in your back yard.

For such an endeavor of plant collecting, a simple plant

press is useful so that specimens may be dried; if this is a hobby about which you would like to know more, I would suggest that you write to the Arnold Arboretum at Jamaica Plain, Mass., for their booklet on botanical collecting.

In order to enjoy nature through collecting plant material, you should certainly use standard illustrated wild flower books of which the new *Field Guide To The Wild Flowers of Northeastern and Northcentral North America* (Peterson and McKenny, Houghton Mifflin & Co., 1974) is possibly the most helpful. When once intrigued with the subject, you could graduate from this to more scientific books like those in the series by H. W. Rickett *Wild Flowers of the United States*, a series that covers the entire country by regions, published by McGraw-Hill, Inc. And, beyond this, for "birding while botanizing," you will need some of the paperback books on birds, on insects, and, for shore use, on fish and on shells.

Perhaps as a beginning camper you do not wish to load yourself down with all the paraphernalia involved in actual collecting. In that case I suggest the camera. A careful "record shot" of a plant or animal may be just as valuable a record as (or sometimes even better than) a brown, dried specimen. I find, after some years of plant collecting in both Mexico and in New England, that my slides are informative to me and to others, and permit me to relive many happy moments.

This is not the place to make detailed recommendations as to what camera you should use or the amount of money to spend. Perhaps the tiny Minox (or similar miniature camera) which I use for the recording of individual flowers is the best answer for the peripatetic camper. Weighing only a few ounces, these cameras are of wide-ranging value.

But the subject of this chapter is camping, and that of this book, the use of wayside plants — so now I would like to suggest how one might meet the needs of a typical day's camping in some "average" wooded or roadside area. Let us take a hypothetical midsummer's day, where the plants you are looking for might be found, although you'll certainly understand that no such ideal situation is ever entirely available.

Let us first take the important matter of food. What kind of menu would use as much as possible of the plant life of the wild?

First, breakfast: one might use the blueberry pickings of the previous day to make delicious blueberry muffins (recipe on p. 33) with coffee, augmented (if you like) with previously dried dandelion roots (p. 39).

A dinner menu might begin with a soup of "greens of the season," or of freshly-gathered mushrooms (pps. 22 & 24) followed by a main dish of meat cooked in grape leaves (p. 28) or, if one is in Florida, meat wrapped in papaya leaves. This would be accompanied by whatever wild vegetable or green salad the season or locale might afford, or possibly sweet potatoes cooked with maple sugar (p. 37). To top off this "wild dinner" there might be that best-of-all-pies, elderberry, which is described on p. 33. Another use of the bounty of the woods would be to collect shavings or chips of hickory wood to broil your meat, thus achieving a flavor akin to the best of smoked meats.

After a meal such as this for dinner, a light supper, by contrast, might be in order. For a meal that's also a conversation piece, and if the season is not too far along, what about a plate of pancakes made from cattail pollen? (p. 34.) Pancakes with maple syrup are our direct heritage from the Indians.

For a nightcap, one might have an easily made, and "oh, so delicious" tiny glass of Chokecherry brandy (p. 39).

But the use of plants for food as suggested is by no means the extent of their usefulness in any day of camping. Besides the various uses suggested in earlier pages of this book, some additional suggestions would be the following:

For mending and tying:

While they are too coarse for actual sewing, the shredded and twisted stems of milkweed or of the marshmallow can easily be made into a crude rope, as strong as the best bought twine.

Sometimes one needs glue rather than string and here one can use quite efficiently the exudations of pine and fir trees or the very gluey globules of exudate on wild cherries. If you don't believe these materials are sticky, try to get them off your hands!

For general cleaning:

For scrubbing pots and pans, a handful of stems of Equisetum will prove really abrasive.

If you have forgotten to bring soap with you, the best substitute you'll find is the plant Saponaria or Soapwort (p. 56).

If there is ironing to be done and if the iron should stick, a few rubs with bayberries will give just the right kind of waxing. One can buy waxing pads made of these berries.

For hobbies:

See Chapters IV and V for things one can do for both fun and practical purposes.

For midday thirsts:

As a real emergency thirst-quencher, I recommend the chewing of a piece of raw onion or, if you are in a woods where wild grapes are growing, you'll find that drops of a safe juice quickly exude from cut stems of such vines. On p. 150 is a recipe for a drink which the Indians told us about, made from the flower heads of the red-flowered sumach.

At night:

In these days of flashlights and gas lamps, the knowledge of the use of cattail stalks, soaked in tallow and used as flares or lamps, is perhaps not so much needed, but this is an easy and cheap way of making your own lamps. Or one could use those same cattail heads, reduced to fluff, to stuff a good camping pillow which might, in an emergency, also serve as a simple life-preserver for a child.

For sickness emergencies in camp:

There are probably few wise campers who would go out in these days without a first aid kit and some aspirin, but it does no harm to know of some of the simple remedies that may be found in the environs of your camp. In more detail these are discussed in Chapter VI of this book, and, at greater length, in my companion volume — *Using Plants For Healing* (Hearthside). But a simple review here might not be out of place.

If nothing else is available, a badly bleeding cut might be safely dressed with squeezed-out, highly absorbent sphagnum moss from a nearby swamp, and as the water in which such moss grows is reputed always to be sterile, this is a valuable bit of knowledge to have for the sudden emergency.

If something poisonous has been eaten, or if vomiting is desirable for whatever reason, a tea made from holly berries and leaves will induce throwing up, as will soap suds similarly swallowed.

For a sore throat, chewing the leaves of the goldthread (Coptis) might bring some relief.

For a cold, if no aspirin is on hand, one might make a steeped tea from the foliage of the sweet-scented goldenrod, adding a bit of Irish moss for its soothing quality, with some lemon juice if you can get it. Honey, from a bee tree if you can find one, or from the store if not, is one of the most healing and healthful of all medicines.

As laxatives, cooked rhubarb, prunes or pears are well-known medicines, as is olive oil, brought along for cooking.

In cases of diarrhoea, the plant most easily found and best for this purpose, is the common blackberry. Tea made from the leaves is effective, as is eating a lot of fruit or drinking blackberry brandy. Other remedies that are variously effective — are an extract of the roots of the common wild geranium, and extracts made from the astringent bark of oaks and hemlocks.

Useful as emergency treatment for burns are such things as a salve of Irish moss and cucumber; an astringent wash of strong tea, or a poultice of bread and hot milk; but most effective of all is just immersing the burned part in cold water.

For toothache, temporary relief can be obtained from the properties of a medicinal plant that unfortunately doesn't grow in America, but which is easily obtained: some oil of cloves.

Flatulence is a not uncommon trouble during summertime camping; it can be alleviated by chewing some leaves of wild mints, or by taking a spoonful of Angostura bitters, if such be available.

Coughs can be eased best with honey or, if you are near a store where it can be bought, some black currant jam.

And of all the troubles most frequently encountered in camping, there is ivy poisoning. There are four different treatments, any one of which should be used as soon as possible after contact with the poison has taken place.

Of these four remedies, the use of the leaves of two common plants is perhaps the easiest and surest. First is the plant often found growing near poison ivy — the jewelweed. The juicy stems and leaves when rubbed on the affected part will help dissolve the volatile oil which is so poisonous. The other plant foliage similarly used is that of the sweet fern, which is often found growing in open barren ground. If neither plant is available, use naphtha soap to wash with, or a drenching with unleaded (white) gasoline. Do not use ordinary high-test gasoline for this purpose.

All of these being but a few of the common uses of wayside plants, it will be seen that there are many valuable properties, uses, and surprises locked in the trees, shrubs, and herbs of our waysides. A little study of these things will make camping more fun, and add to your comfort and health.

BIBLIOGRAPHY

(Revised 1979)

Adrosko, Rita J. *Natural Dyes and Home Dyeing*. New York: Dover Publications, 1971.

Aikman, L. *Nature's Healing Arts, From Folk Medicine to Modern Drugs*. Washington, D.C.: National Geographic, 1977.

Allegro, John M. *The Sacred Mushroom and the Cross*. New York: Doubleday & Co., 1970.

Altschul, Sir von Ries. *Drugs and Foods from Little Known Plants*. Cambridge, Mass.: Harvard University Press, 1973.

Angier, Bradford. *Free for the Eating*. Harrisburg, Pa.: Stackpole Co., 1966.

———. *Living off the Country*. Harrisburg, Pa.: Stackpole Co., 1961.

———. *More Free for the Eating Wild Foods*. Harrisburg, Pa.: Stackpole Co., 1969.

———. *How to Live in the Woods on Pennies a Day*. Harrisburg, Pa.: Stackpole Co., 1971.

———. *Wilderness Cookery*. Harrisburg, Pa.: Stackpole Co., 1970.

Arnold, Harry L. *Poisonous Plants of Hawaii*. Rutland, Vt.: Charles E. Tuttle Co., 1968.

Badianus. *Badianus Manuscript*. Baltimore, Maryland: The Johns Hopkins Press, 1940.

Bailey, Liberty H. *Hortus Second*. New York: Macmillan, 1958.

Balls, E. K. *Early Use of California Plants*. Berkeley, Ca.: University of California Press, 1965.

Barton, Benjamin Smith. *Collections for an Essay Towards a Materia Medica of the United States*. Philadelphia, 1798 and 1804.

Beauchamp, W. M. *Aboriginal Use of Wood in New York*. New York: AMS Press, 1977.

Belanger, Emil J. *Modern Manufacturing Formulary*. New York: Chemical Publishing Co., 1958.

Bemiss, Elijah. *The Dyer's Companion.* New York: Dover Publications, 1973.

Berglund, Berndt. *The Edible Wild.* New York: Charles Scribner's Sons, 1971.

Berglund, Berndt and Bolsby, Clare E. *The Complete Outdoorsman's Guide to Edible Wild Plants.* New York: Charles Scribner's Sons, 1978.

Blair, Thos. S. *Botanic Drugs: Their Materia Medica, Pharmacology and Therapeutics.* New York: Gordon & Breach, Science Publishers, 1976.

Blunt, Wilfrid. *Art of Botanical Illustration.* London: William Collins, 1950.

Brimer, John. *Growing Herbs in Pots.* New York: Simon & Schuster, 1976.

Brown, O. Phelps. *The Complete Herbalist.* Jersey City, 1856.

Buchman, Dian D. *Complete Herbal Guide to Natural Health & Beauty.* New York: Doubleday & Co., 1973.

Budge, E. A. Wallis. *Divine Origin of the Craft of the Herbalist.* Detroit, Mich.: Gale Research Company, 1971.

Burn, J. Harold. *Drugs, Medecine and Man.* New York: Charles Scribner's Sons, 1962.

Castleton, Virginia. *Handbook of Natural Beauty.* Emmaus, Pa.: Rodale Press, 1975.

Christensen, Clyde M. *Common Edible Mushrooms.* Minneapolis, Minn.: University of Minnesota Press, 1969.

Clark, Linda. *Get Well Naturally.* Old Greenwich, Conn.: Devin-Adair, 1974.

Clarkson, Rosetta E. *The Golden Age of Herbs and Herbalists.* New York: Dover Publications, 1972.

Claus, Edw. P. and Tyler, Varro E., Jr. *Pharmacognosy.* Philadelphia, Pa.: Lea & Febiger, 1970.

Coon, Nelson. *Using Plants for Healing.* Emmaus, Pa.: Rodale Press, 1979.

———. *Dictionary of Useful Plants.* Emmaus, Pa.: Rodale Press, 1977.

Craighead, John J. et al. *A Field Guide to Rocky Mountain Wild Flowers.* Boston, Mass.: Houghton Mifflin Co., 1974.

Creekmore, Betsey B. *Traditional American Crafts: A Practical Guide to 300 Years Methods & Materials.* New York: Hearthside Press, 1970.

Creekmore, Hubert. *Daffodils Are Dangerous: The Poisonous Plants in Your Garden.* New York: Walker & Co., 1966.

Crowhurst, Adrian. *The Weed Cookbook.* New York: Lancer Books, 1972.

Culpeper, Nicholas. *The Complete Herbal and English Physician Enlarged.* London, 1653.

Curtin, L. S. M. *Healing Herbs of the Upper Rio Grande.* Santa Fe, N. Mex.: Santa Fe Laboratory of Anthropology, 1947.

Davidson, Mary F. *The Dye Pot.* M. F. Davidson, 1974.

de Bairacli-Levy, Juliette. *Herbal Handbook for Farm.* Emmaus, Pa.: Rodale Press, 1976.

———. *Herbal Handbook for Everyone.* Newton Centre, Mass.: Branford, 1966.

———. *Nature's Children: A Guide to Organic Foods and Herbal Remedies for Children.* New York: Schocken Books, 1978.

———. *Complete Herbal Book for the Dog.* New York: Arco Publishing Co., 1972.

de Candolle, A. L. P. *Origin of Cultivated Plants.* New York: Hafner Publishing Co., 1967.

de Laszlo, Henry G. *Library of Medicinal Plants.* Cambridge, England: Hefner and Sons, 1958.

Dioscorides, Pedanius. *Greek Herbal.* New York: Hafner Publishing Co., 1968.

Fenton, William N. *Contacts between Iroquois Herbalism and Colonial Medicine.* Washington, D.C.: Smithsonian Institute, 1941–1942.

Fernald, Merritt L. et al. *Edible Wild Plants of Eastern North America.* New York: Harper & Row Publishers, 1958.

Fernie, W. T. *Herbal Simples.* London: Simpkin, 1914.

Findlay, W. P. *Wayside and Woodland Fungi.* London: Frederick Warne, 1967.

Foster, Gertrude B. *Herbs for Every Garden.* New York: Dutton, 1973.

Freeman, Margaret B. *Herbs for the Mediaeval Household, for Cooking, Healing, and Divers Uses.* New York: Metropolitan Museum of Art, 1943.

Gerard, John. *The Herbal or General History of Plants.* New York: Dover Publications, 1975.

Gibbons, Euell. *Stalking the Healthful Herbs.* New York: David McKay Co., 1970.

———. *Stalking the Wild Asparagus.* New York: David McKay Co., 1970.

Gilkey, Helen. *Handbook of Northwest Flowering Plants.* Portland, Ore.: Binford and Mort Publishers, 1951.

Gillespie, W. H. *A Compilation of the Edible Wild Plants of West Virginia.* New York: Press of Scholar's Library, 1959.

Gilmore, Melvin R. *Uses of Plants by the Indians of the Missouri River Region.* Lincoln, Nebraska: University of Nebraska Press, 1977.

Gouzil, Dezerina. *Mother Nature's Herbs & Teas.* New York: Oliver Press, 1975.

Graham, Verne O. *Mushrooms of the Great Lakes Area.* New York: Dover Publications, 1973.

Gray, Asa. *Manual of Botany.* New York: American Book Company, 1950.

Grieve, Maude. *A Modern Herbal.* New York: Dover Publications, 1971.

Grigson, Geoffrey. *The Englishman's Flora.* London: Phoenix House, 1955.

Gunther, Erna. *Ethnobotany of Western Washington: The Knowledge and Use of Indigenous Plants by Native Americans.* Seattle, Wash.: University of Washington, Press, 1973.

Hall, Dorothy. *The Book of Herbs.* New York: Charles Scribner's Sons, 1974.

Hardacre, Val. *Woodland Nuggets of Gold-Ginseng.* New York: Vantage Press, 1968.

Hardin, James W. and Arena, James W. *Human Poisoning from Native and Cultivated Plants.* Durham, N. C.: Duke University Press, 1973.

Hariot, Thomas. *The First English Plantation of Virginia.* London, 1588.

Harrington, H. D. *Edible Native Plants of the Rocky Mountains.* Albuquerque, N. Mex.: University of New Mexico Press, 1974.

Harris, Ben Charles. *Eat the Weeds.* Barre, Mass.: Barre Publishers, 1968.

———. *Kitchen Medicines.* Barre, Mass.: Barre Publishers, 1970.

Haskin, Leslie. *Wildflowers of the Pacific Northwest.* Portland, Oreg.: Binford and Mort Publishers.

Hatfield, Audrey W. *How To Enjoy Your Weeds.* New York: Macmillan, 1973.

———. *Pleasures of Wild Plants.* New York: Taplinger, 1967.

Hatton, Richard G. *Handbook of Plant & Floral Ornament.* New York: Dover Publications, 1960.

Hausman, E. H. *Beginner's Guide to Wild Flowers.* New York: G. P. Putnam and Sons, 1948.

Hedrick, U. P. *Sturtevant's Edible Plants of the World.* New York: Dover Publications, 1972.

Hill, Albert F. *Economic Botany.* New York: McGraw-Hill Book Co., 1952.

Hunter, Beatrice T. *Gardening Without Poison.* Boston, Mass.: Houghton Mifflin Co., 1972.

Jacob, Dorothy. *Witch's Guide to Gardening.* New York: Taplinger Publishing Co., 1966.

Jaeger, Ellsworth. *Easy Crafts.* New York: Macmillan, 1947.

———. *Nature Crafts.* New York: Macmillan, 1950.

Jarvis, D. C. *Folk Medicine.* New York: Henry Holt and Co., Inc., 1958.

Johnson, C. P. *The Useful Plants of Great Britain.* London: Hardwicke, 1861.

Johnson, J. R. *Anyone Can Live off the Land.* New York: David McKay Co., 1961.

Josselyn, John. *New England Rarities Discovered.* London, 1672.

Kadans, Joseph. *Encyclopedia of Medicinal Herbs.* New York: Arc Books, 1972.

Kephart, Horace. *Camping and Woodcraft.* New York: Macmillan, 1948.

Kimbal, Jeffe and Anderson, Jean. *The Art of American Indian Cooking.* New York: Doubleday & Co., 1965.

Kingsbury, John M. *Poisonous Plants of the United States and Canada.* Englewood Cliffs, N. J.: Prentice-Hall, 1964.

———. *Deadly Harvest: A Guide To Common Poisonous Plants.* New York: Holt, Rinehart & Winston, Inc., 1965.

Kierstead, Sallie P. *Natural Dyes.* Boston, Mass.: Branden Press, 1950.

Kirk, Donald R. *Wild Edible Plants of the Western United States.* Happy Camp, Calif.: Naturegraph Publishers, 1970.

Kramer, Jack. *Natural Dyes: Plants and Processes.* New York: Charles Scribner's Sons, 1972.

Krieger, Louis C. *Mushroom Handbook.* New York: Dover Publications, 1967.

Krochmal, Arnold and Krochmal, Connie. *A Guide to Medicinal Plants of the United States.* New York: New York Times Books, 1973.

Law, Donald. *Herb Growing for Health.* New York: Arc Books, 1969.

Lee, Charles A. *Medicinal Plants Growing in the State of New York.* New York: Langley, 1848.

Leighton, Ann. *Early American Gardens for "Meate and Medecine".* Boston, Mass.: Houghton Mifflin Co., 1970.

Lesch, Alma. *Vegetable Dyeing.* New York: Watson-Gupthill Publications, 1971.

Leyel, Mrs. C. F. *Elixirs of Life.* London: Faber & Faber, 1958.

———. *Green Medicine.* London: Faber & Faber, 1952.

Lindley, John. *Medical and Economic Botany.* London: Bradbury and Evans, 1856.

Lucas, Richard. *Common and Uncommon Uses for Herbs for Healthful Living.* New York: Arc Books, 1970.

McIlvaine, Charles and Macadam, Robert. *One Thousand American Fungi.* New York: Dover Publications, 1973.

MacNicol, Mary. *Flower Cookery.* New York: Macmillan, 1972.

Marshall, Humphry. *Arbustum Americanum.* New York: Hafner Publishing Co., 1967.

Mathews, F. Schuyler. *Field Book of American Wild Flowers.* New York: G. P. Putnam and Sons, 1915.
Medsger, Oliver P. *Edible Wild Plants.* New York: Macmillan, 1972.
Meehan, Thomas. *The Native Wild Flowers and Ferns of the United States.* Boston: Prang, 1878.
Messegue, Maurice. *Of Men and Plants.* New York: Macmillan, 1973.
Meyer, Joseph. *Herbalist.* New York: Sterling Publishing Co., 1968.
Millspaugh, Charles F. *American Medicinal Plants.* New York: Dover Publications, 1974.
Morton, Julia. *Wild Plants for Survival in South Florida.* Miami, Florida: Hurricane House, 1963.
Muenscher, Walter C. *Poisonous Plants of the United States.* New York: Macmillan, 1975.
Muenscher, Walter C. and Rice, Myron C. *Garden Spice and Wild Pot-Herbs.* Ithaca, N. Y.: Cornell University Press, 1978.
Ormond, Clyde. *Complete Book of Outdoor Lore.* New York: Harper and Row Publishers, 1965.
Petrides, George A. *Field Guide to Trees and Shrubs.* Boston, Mass: Houghton Mifflin Co., 1973.
Quelch, Mary Thorne. *Herbs for Daily Use in Home Medicine and Cookery.* London: Faber & Faber, 1941.
———. *The Herb Garden.* London: Faber & Faber, 1941.
Revolutionary Health Committee of Hunan Province. *A Barefoot Doctor's Manual.* Seattle, Wash.: Cloudburst Press, 1977.
Rickett, Harold W. *Wildflowers of the United States.* New York: McGraw-Hill, Inc., 1966.
Robbins, Wilfred W. and Ramaley, Francis. *Plants Useful to Man.* New York: Blakiston's Sons & Co., 1937.
Romero, John B. *Botanical Lore of California Indians.* New York: Vantage Press, 1954.
Saunders, Charles F. *Edible and Useful Plants of the United States and Canada.* New York: Dover Publications, 1934.
Schafer, Violet. *Herbcraft.* Bribane, Calif.: Taylor & NG, 1971.
Schaffer, Florence M. *Driftwood Miniatures.* New York: Hearthside Press, 1967.
Scully, Virginia. *Treasury of American Indian Herbs: Their Lore & Their Use for Food and Medicine.* New York: Crown Publishers, 1970.
Simmonite, W. J. and Culpeper, N. *The Simmonite-Culpeper Herbal Remedies.* New York: Sterling Publishing Co., 1957.
Simmons, Adelma G. *Herb Gardening in Five Seasons.* New York: Hawthorn Press, 1977.

Smith, Alexander H. *Mushroom Hunter's Field Guide.* Ann Arbor, Mich.: University of Michigan Press, 1963.

Snell, Walter and Duck, Esther A. *A Glossary of Mycology.* Boston, Mass.: Harvard University Press, 1971.

Spencer, E. R. *All About Weeds.* New York: Dover Publications, 1974.

Squires, Mabel. *New Trends in Dried Arrangements and Decorations.* Woodbury, N. Y.: Barron's Educational Series, 1967.

Stone, Eric P. *Medicine Among the American Indians.* New York: AMS Press, 1978.

Sweet, Muriel. *Common Edible & Useful Plants of the West.* Happy Camp, Calif.: Naturegraph Publishers, 1962.

Taylor, Kathryn S. *A Traveler's Guide to Roadside Wildflowers, Shrubs and Trees of the U. S.* New York: Farrar, Strauss & Giroux, Inc., 1949.

Taylor, Lyda A. *Plants Used as Curatives by Certain Southeastern Tribes.* New York: AMS Press, 1977.

Tenebaum, Frances. *Gardening with Wildflowers.* New York: Charles Scribner's Sons, 1973.

Theophrastus. *Enquiry into Plants.* Boston, Mass.: Harvard University Press.

Thomas, Mai, ed. *Grannies' Remedies.* New York: Crown Publishers, 1967.

Thomas, William S. *Field Book of Common Mushrooms.* New York: G. P. Putnam's Sons, 1948.

Thompson, Dorothea. *Creative Decorations with Dried Flowers.* New York: Hearthside Press, 1972.

Tisserand, Robert B. *The Art of Aromatherapy: A Handbook of the Beautifying & Healing Properties of the Essential Oils of Flowers and Herbs.* New York: Inner Traditions International, 1978.

Uphof, J. C. *Dictionary of Economic Plants.* Lubrecht & Cramer, 1968.

Von Miklos, Josephine. *Wild Flowers in Your House.* New York: Doubleday & Co., 1968.

Wakefield, E. M. and Dennis, R. W. G. *Common British Fungi.* London: Gawthorn, 1950.

Wasson, R. Gordon. *Soma, the Divine Mushroom of Immortality.* New York: Harcourt Brace, 1971.

Webster, Helen N. *Herbs, How to Grow Them and How to Use Them.* Newton Centre, Mass.: Charles T. Branford Co.

Weiner, Michael. *Earth Medicine—Earth Foods: Plant Remedies, Drugs and Natural Foods of the North American Indians.* New York: Macmillan, 1972.

Wherry, Edgar T. *Wild Flower Guide.* New York: Doubleday & Co., 1948.

Whitlock, Sarah and Rankin, Martha. *New Techniques with Dried Flowers.* New York: Dover Publications, 1962.

Wigginton, Eliot. *Foxfire Book.* New York: Doubleday & Co., 1975.

Wilder, Walter Beebe. *Bounty of the Wayside.* New York: Doubleday & Co., 1946.

Woodward, Marcus. *Leaves from Gerard's Herbal.* New York: Dover Publications, 1969.

Wren, R. C. *Potter's New Cyclopedia of Botanical Drugs and Preparations.* London: Pitman, 1956.

Youngken, Heber W. *Textbook of Pharmacognosy.* New York: McGraw-Hill, 1948.

INDEX

The bold-faced numerals refer to the principal discussion of the plant found in Section 2, while more detailed uses of the plants are found on pages shown in light-faced numerals. Colloquial names bear references to the principal listing of the scientific name.

A CATALOGUE OF SELECTED DOVER BOOKS IN ALL FIELDS OF INTEREST

A CATALOGUE OF SELECTED DOVER BOOKS IN ALL FIELDS OF INTEREST

CELESTIAL OBJECTS FOR COMMON TELESCOPES, T. W. Webb. The most used book in amateur astronomy: inestimable aid for locating and identifying nearly 4,000 celestial objects. Edited, updated by Margaret W. Mayall. 77 illustrations. Total of 645pp. 5⅜ x 8½.
20917-2, 20918-0 Pa., Two-vol. set **$10.00**

HISTORICAL STUDIES IN THE LANGUAGE OF CHEMISTRY, M. P. Crosland. The important part language has played in the development of chemistry from the symbolism of alchemy to the adoption of systematic nomenclature in 1892. ". . . wholeheartedly recommended,"—Science. 15 illustrations. 416pp. of text. 5⅝ x 8¼. 63702-6 Pa. **$7.50**

BURNHAM'S CELESTIAL HANDBOOK, Robert Burnham, Jr. Thorough, readable guide to the stars beyond our solar system. Exhaustive treatment, fully illustrated. Breakdown is alphabetical by constellation: Andromeda to Cetus in Vol. 1; Chamaeleon to Orion in Vol. 2; and Pavo to Vulpecula in Vol. 3. Hundreds of illustrations. Total of about 2000pp. 6⅛ x 9¼.
23567-X, 23568-8, 23673-0 Pa., Three-vol. set $32.85

THEORY OF WING SECTIONS: INCLUDING A SUMMARY OF AIRFOIL DATA, Ira H. Abbott and A. E. von Doenhoff. Concise compilation of subatomic aerodynamic characteristics of modern NASA wing sections, plus description of theory. 350pp. of tables. 693pp. 5⅜ x 8½.
60586-8 Pa. **$9.95**

DE RE METALLICA, Georgius Agricola. Translated by Herbert C. Hoover and Lou H. Hoover. The famous Hoover translation of greatest treatise on technological chemistry, engineering, geology, mining of early modern times (1556). All 289 original woodcuts. 638pp. 6¾ x 11.
60006-8 Clothbd. **$19.95**

THE ORIGIN OF CONTINENTS AND OCEANS, Alfred Wegener. One of the most influential, most controversial books in science, the classic statement for continental drift. Full 1966 translation of Wegener's final (1929) version. 64 illustrations. 246pp. 5⅜ x 8½.(EBE)61708-4 Pa. **$5.00**

THE PRINCIPLES OF PSYCHOLOGY, William James. Famous long course complete, unabridged. Stream of thought, time perception, memory, experimental methods; great work decades ahead of its time. Still valid, useful; read in many classes. 94 figures. Total of 1391pp. 5⅜ x 8½.
20381-6, 20382-4 Pa., Two-vol. set **$19.90**

THE SENSE OF BEAUTY, George Santayana. Masterfully written discussion of nature of beauty, materials of beauty, form, expression; art, literature, social sciences all involved. 168pp. 5⅜ x 8½. 20238-0 Pa. **$3.50**

ON THE IMPROVEMENT OF THE UNDERSTANDING, Benedict Spinoza. Also contains *Ethics, Correspondence,* all in excellent R. Elwes translation. Basic works on entry to philosophy, pantheism, exchange of ideas with great contemporaries. 402pp. 5⅜ x 8½. 20250-X Pa. $5.95

THE TRAGIC SENSE OF LIFE, Miguel de Unamuno. Acknowledged masterpiece of existential literature, one of most important books of 20th century. Introduction by Madariaga. 367pp. 5⅜ x 8½.
20257-7 Pa. $6.00

THE GUIDE FOR THE PERPLEXED, Moses Maimonides. Great classic of medieval Judaism attempts to reconcile revealed religion (Pentateuch, commentaries) with Aristotelian philosophy. Important historically, still relevant in problems. Unabridged Friedlander translation. Total of 473pp. 5⅜ x 8½. 20351-4 Pa. $6.95

THE I CHING (THE BOOK OF CHANGES), translated by James Legge. Complete translation of basic text plus appendices by Confucius, and Chinese commentary of most penetrating divination manual ever prepared. Indispensable to study of early Oriental civilizations, to modern inquiring reader. 448pp. 5⅜ x 8½. 21062-6 Pa. $6.00

THE EGYPTIAN BOOK OF THE DEAD, E. A. Wallis Budge. Complete reproduction of Ani's papyrus, finest ever found. Full hieroglyphic text, interlinear transliteration, word for word translation, smooth translation. Basic work, for Egyptology, for modern study of psychic matters. Total of 533pp. 6½ x 9¼. (USCO) 21866-X Pa. $8.50

THE GODS OF THE EGYPTIANS, E. A. Wallis Budge. Never excelled for richness, fullness: all gods, goddesses, demons, mythical figures of Ancient Egypt; their legends, rites, incarnations, variations, powers, etc. Many hieroglyphic texts cited. Over 225 illustrations, plus 6 color plates. Total of 988pp. 6⅛ x 9¼. (EBE)
22055-9, 22056-7 Pa., Two-vol. set $20.00

THE STANDARD BOOK OF QUILT MAKING AND COLLECTING, Marguerite Ickis. Full information, full-sized patterns for making 46 traditional quilts, also 150 other patterns. Quilted cloths, lame, satin quilts, etc. 483 illustrations. 273pp. 6⅞ x 9⅝. 20582-7 Pa. $5.95

CORAL GARDENS AND THEIR MAGIC, Bronsilaw Malinowski. Classic study of the methods of tilling the soil and of agricultural rites in the Trobriand Islands of Melanesia. Author is one of the most important figures in the field of modern social anthropology. 143 illustrations. Indexes. Total of 911pp. of text. 5⅝ x 8¼. (Available in U.S. only)
23597-1 Pa. $12.95

THE AMERICAN SENATOR, Anthony Trollope. Little known, long unavailable Trollope novel on a grand scale. Here are humorous comment on American vs. English culture, and stunning portrayal of a heroine/villainess. Superb evocation of Victorian village life. 561pp. 5⅜ x 8½.
23801-6 Pa. **$7.95**

WAS IT MURDER? James Hilton. The author of *Lost Horizon* and *Goodbye, Mr. Chips* wrote one detective novel (under a pen-name) which was quickly forgotten and virtually lost, even at the height of Hilton's fame. This edition brings it back—a finely crafted public school puzzle resplendent with Hilton's stylish atmosphere. A thoroughly English thriller by the creator of Shangri-la. 252pp. 5⅜ x 8. (Available in U.S. only)
23774-5 Pa. $3.00

CENTRAL PARK: A PHOTOGRAPHIC GUIDE, Victor Laredo and Henry Hope Reed. 121 superb photographs show dramatic views of Central Park: Bethesda Fountain, Cleopatra's Needle, Sheep Meadow, the Blockhouse, plus people engaged in many park activities: ice skating, bike riding, etc. Captions by former Curator of Central Park, Henry Hope Reed, provide historical view, changes, etc. Also photos of N.Y. landmarks on park's periphery. 96pp. 8½ x 11. 23750-8 Pa. **$4.95**

NANTUCKET IN THE NINETEENTH CENTURY, Clay Lancaster. 180 rare photographs, stereographs, maps, drawings and floor plans recreate unique American island society. Authentic scenes of shipwreck, lighthouses, streets, homes are arranged in geographic sequence to provide walking-tour guide to old Nantucket existing today. Introduction, captions. 160pp. 8⅞ x 11¾. 23747-8 Pa. **$7.95**

STONE AND MAN: A PHOTOGRAPHIC EXPLORATION, Andreas Feininger. 106 photographs by *Life* photographer Feininger portray man's deep passion for stone through the ages. Stonehenge-like megaliths, fortified towns, sculpted marble and crumbling tenements show textures, beauties, fascination. 128pp. 9¼ x 10¾. 23756-7 Pa. **$6.95**

CIRCLES, A MATHEMATICAL VIEW, D. Pedoe. Fundamental aspects of college geometry, non-Euclidean geometry, and other branches of mathematics: representing circle by point. Poincare model, isoperimetric property, etc. Stimulating recreational reading. 66 figures. 96pp. 5⅝ x 8¼.
63698-4 Pa. **$3.50**

THE DISCOVERY OF NEPTUNE, Morton Grosser. Dramatic scientific history of the investigations leading up to the actual discovery of the eighth planet of our solar system. Lucid, well-researched book by well-known historian of science. 172pp. 5⅜ x 8½. 23726-5 Pa. **$3.95**

THE DEVIL'S DICTIONARY. Ambrose Bierce. Barbed, bitter, brilliant witticisms in the form of a dictionary. Best, most ferocious satire America has produced. 145pp. 5⅜ x 8½. 20487-1 Pa. $2.50

THE ART OF THE CINEMATOGRAPHER, Leonard Maltin. Survey of American cinematography history and anecdotal interviews with 5 masters—Arthur Miller, Hal Mohr, Hal Rosson, Lucien Ballard, and Conrad Hall. Very large selection of behind-the-scenes production photos. 105 photographs. Filmographies. Index. Originally *Behind the Camera.* 144pp. 8¼ x 11. 23686-2 Pa. $5.00

THE COMPLETE NONSENSE OF EDWARD LEAR, Edward Lear. All nonsense limericks, zany alphabets, Owl and Pussycat, songs, nonsense botany, etc., illustrated by Lear. Total of 321pp. 5⅜ x 8½. (Available in U.S. only) 20167-8 Pa. $4.50

INGENIOUS MATHEMATICAL PROBLEMS AND METHODS, Louis A. Graham. Sophisticated material from Graham *Dial,* applied and pure; stresses solution methods. Logic, number theory, networks, inversions, etc. 237pp. 5⅜ x 8½. 20545-2 Pa. $4.95

BEST MATHEMATICAL PUZZLES OF SAM LOYD, edited by Martin Gardner. Bizarre, original, whimsical puzzles by America's greatest puzzler. From fabulously rare *Cyclopedia,* including famous 14-15 puzzles, the Horse of a Different Color, 115 more. Elementary math. 150 illustrations. 167pp. 5⅜ x 8½. 20498-7 Pa. $3.50

THE BASIS OF COMBINATION IN CHESS, J. du Mont. Easy-to-follow, instructive book on elements of combination play, with chapters on each piece and every powerful combination team—two knights, bishop and knight, rook and bishop, etc. 250 diagrams. 218pp. 5⅜ x 8½. (Available in U.S. only) 23644-7 Pa. $4.50

MODERN CHESS STRATEGY, Ludek Pachman. The use of the queen, the active king, exchanges, pawn play, the center, weak squares, etc. Section on rook alone worth price of the book. Stress on the moderns. Often considered the most important book on strategy. 314pp. 5⅜ x 8½. 20290-9 Pa. $5.00

LASKER'S MANUAL OF CHESS, Dr. Emanuel Lasker. Great world champion offers very thorough coverage of all aspects of chess. Combinations, position play, openings, end game, aesthetics of chess, philosophy of struggle, much more. Filled with analyzed games. 390pp. 5⅜ x 8½. 20640-8 Pa. $5.95

500 MASTER GAMES OF CHESS, S. Tartakower, J. du Mont. Vast collection of great chess games from 1798-1938, with much material nowhere else readily available. Fully annoted, arranged by opening for easier study. 664pp. 5⅜ x 8½. 23208-5 Pa. $8.50

A GUIDE TO CHESS ENDINGS, Dr. Max Euwe, David Hooper. One of the finest modern works on chess endings. Thorough analysis of the most frequently encountered endings by former world champion. 331 examples, each with diagram. 248pp. 5⅜ x 8½. 23332-4 Pa. $3.95

THE COMPLETE BOOK OF DOLL MAKING AND COLLECTING, Catherine Christopher. Instructions, patterns for dozens of dolls, from rag doll on up to elaborate, historically accurate figures. Mould faces, sew clothing, make doll houses, etc. Also collecting information. Many illustrations. 288pp. 6 x 9. 22066-4 Pa. **$4.95**

THE DAGUERREOTYPE IN AMERICA, Beaumont Newhall. Wonderful portraits, 1850's townscapes, landscapes; full text plus 104 photographs. The basic book. Enlarged 1976 edition. 272pp. 8¼ x 11¼. 23322-7 Pa. $7.95

CRAFTSMAN HOMES, Gustav Stickley. 296 architectural drawings, floor plans, and photographs illustrate 40 different kinds of "Mission-style" homes from *The Craftsman* (1901-16), voice of American style of simplicity and organic harmony. Thorough coverage of Craftsman idea in text and picture, now collector's item. 224pp. 8⅛ x 11. 23791-5 Pa. **$6.50**

PEWTER-WORKING: INSTRUCTIONS AND PROJECTS, Burl N. Osborn. & Gordon O. Wilber. Introduction to pewter-working for amateur craftsman. History and characteristics of pewter; tools, materials, step-by-step instructions. Photos, line drawings, diagrams. Total of 160pp. 7⅞ x 10¾. 23786-9 Pa. **$4.50**

THE GREAT CHICAGO FIRE, edited by David Lowe. 10 dramatic, eyewitness accounts of the 1871 disaster, including one of the aftermath and rebuilding, plus 70 contemporary photographs and illustrations of the ruins—courthouse, Palmer House, Great Central Depot, etc. Introduction by David Lowe. 87pp. 8¼ x 11. 23771-0 Pa. **$4.95**

SILHOUETTES: A PICTORIAL ARCHIVE OF VARIED ILLUSTRATIONS, edited by Carol Belanger Grafton. Over 600 silhouettes from the 18th to 20th centuries include profiles and full figures of men and women, children, birds and animals, groups and scenes, nature, ships, an alphabet. Dozens of uses for commercial artists and craftspeople. 144pp. 8⅜ x 11¼. 23781-8 Pa. $4.50

ANIMALS: 1,419 COPYRIGHT-FREE ILLUSTRATIONS OF MAMMALS, BIRDS, FISH, INSECTS, ETC., edited by Jim Harter. Clear wood engravings present, in extremely lifelike poses, over 1,000 species of animals. One of the most extensive copyright-free pictorial sourcebooks of its kind. Captions. Index. 284pp. 9 x 12. 23766-4 Pa. $8.95

INDIAN DESIGNS FROM ANCIENT ECUADOR, Frederick W. Shaffer. 282 original designs by pre-Columbian Indians of Ecuador (500-1500 A.D.). Designs include people, mammals, birds, reptiles, fish, plants, heads, geometric designs. Use as is or alter for advertising, textiles, leathercraft, etc. Introduction. 95pp. 8¾ x 11¼. 23764-8 Pa. **$4.95**

SZIGETI ON THE VIOLIN, Joseph Szigeti. Genial, loosely structured tour by premier violinist, featuring a pleasant mixture of reminiscenes, insights into great music and musicians, innumerable tips for practicing violinists. 385 musical passages. 256pp. 5⅝ x 8¼. 23763-X Pa. **$5.00**

TONE POEMS, SERIES II: TILL EULENSPIEGELS LUSTIGE STREICHE, ALSO SPRACH ZARATHUSTRA, AND EIN HELDEN-LEBEN, Richard Strauss. Three important orchestral works, including very popular *Till Eulenspiegel's Marry Pranks,* reproduced in full score from original editions. Study score. 315pp. 9⅜ x 12¼. (Available in U.S. only)
23755-9 Pa. $9.95

TONE POEMS, SERIES I: DON JUAN, TOD UND VERKLARUNG AND DON QUIXOTE, Richard Strauss. Three of the most often performed and recorded works in entire orchestral repertoire, reproduced in full score from original editions. Study score. 286pp. 9⅜ x 12¼. (Available in U.S. only)
23754-0 Pa. $9.95

11 LATE STRING QUARTETS, Franz Joseph Haydn. The form which Haydn defined and "brought to perfection." *(Grove's).* 11 string quartets in complete score, his last and his best. The first in a projected series of the complete Haydn string quartets. Reliable modern Eulenberg edition, otherwise difficult to obtain. 320pp. 8⅜ x 11¼. (Available in U.S. only)
23753-2 Pa. $8.95

FOURTH, FIFTH AND SIXTH SYMPHONIES IN FULL SCORE, Peter Ilyitch Tchaikovsky. Complete orchestral scores of Symphony No. 4 in F Minor, Op. 36; Symphony No. 5 in E Minor, Op. 64; Symphony No. 6 in B Minor, "Pathetique," Op. 74. Bretikopf & Hartel eds. Study score. 480pp. 9⅜ x 12¼.
23861-X Pa. $12.95

THE MARRIAGE OF FIGARO: COMPLETE SCORE, Wolfgang A. Mozart. Finest comic opera ever written. Full score, not to be confused with piano renderings. Peters edition. Study score. 448pp. 9⅜ x 12¼. (Available in U.S. only)
23751-6 Pa. $13.95

"IMAGE" ON THE ART AND EVOLUTION OF THE FILM, edited by Marshall Deutelbaum. Pioneering book brings together for first time 38 groundbreaking articles on early silent films from *Image* and 263 illustrations newly shot from rare prints in the collection of the International Museum of Photography. A landmark work. Index. 256pp. 8¼ x 11.
23777-X Pa. $8.95

AROUND-THE-WORLD COOKY BOOK, Lois Lintner Sumption and Marguerite Lintner Ashbrook. 373 cooky and frosting recipes from 28 countries (America, Austria, China, Russia, Italy, etc.) include Viennese kisses, rice wafers, London strips, lady fingers, hony, sugar spice, maple cookies, etc. Clear instructions. All tested. 38 drawings. 182pp. 5⅜ x 8.
23802-4 Pa. $2.75

THE ART NOUVEAU STYLE, edited by Roberta Waddell. 579 rare photographs, not available elsewhere, of works in jewelry, metalwork, glass, ceramics, textiles, architecture and furniture by 175 artists—Mucha, Seguy, Lalique, Tiffany, Gaudin, Hohlwein, Saarinen, and many others. 288pp. 8⅜ x 11¼.
23515-7 Pa. $8.95

THE CURVES OF LIFE, Theodore A. Cook. Examination of shells, leaves, horns, human body, art, etc., in "*the* classic reference on how the golden ratio applies to spirals and helices in nature . . . ."—Martin Gardner. 426 illustrations. Total of 512pp. 5⅜ x 8½. 23701-X Pa. **$6.95**

AN ILLUSTRATED FLORA OF THE NORTHERN UNITED STATES AND CANADA, Nathaniel L. Britton, Addison Brown. Encyclopedic work covers 4666 species, ferns on up. Everything. Full botanical information, illustration for each. This earlier edition is preferred by many to more recent revisions. 1913 edition. Over 4000 illustrations, total of 2087pp. 6⅛ x 9¼. 22642-5, 22643-3, 22644-1 Pa., Three-vol. set **$28.50**

MANUAL OF THE GRASSES OF THE UNITED STATES, A. S. Hitchcock, U.S. Dept. of Agriculture. The basic study of American grasses, both indigenous and escapes, cultivated and wild. Over 1400 species. Full descriptions, information. Over 1100 maps, illustrations. Total of 1051pp. 5⅜ x 8½. 22717-0, 22718-9 Pa., Two-vol. set **$17.00**

THE CACTACEAE,, Nathaniel L. Britton, John N. Rose. Exhaustive, definitive. Every cactus in the world. Full botanical descriptions. Thorough statement of nomenclatures, habitat, detailed finding keys. The one book needed by every cactus enthusiast. Over 1275 illustrations. Total of 1080pp. 8 x 10¼. 21191-6, 21192-4 Clothbd., Two-vol. set **$50.00**

AMERICAN MEDICINAL PLANTS, Charles F. Millspaugh. Full descriptions, 180 plants covered: history; physical description; methods of preparation with all chemical constituents extracted; all claimed curative or adverse effects. 180 full-page plates. Classification table. 804pp. 6½ x 9¼. 23034-1 Pa. **$13.95**

A MODERN HERBAL, Margaret Grieve. Much the fullest, most exact, most useful compilation of herbal material. Gigantic alphabetical encyclopedia, from aconite to zedoary, gives botanical information, medical properties, folklore, economic uses, and much else. Indispensable to serious reader. 161 illustrations. 888pp. 6½ x 9¼. (Available in U.S. only) 22798-7, 22799-5 Pa., Two-vol. set **$15.00**

THE HERBAL or GENERAL HISTORY OF PLANTS, John Gerard. The 1633 edition revised and enlarged by Thomas Johnson. Containing almost 2850 plant descriptions and 2705 superb illustrations, Gerard's *Herbal* is a monumental work, the book all modern English herbals are derived from, the one herbal every serious enthusiast should have in its entirety. Original editions are worth perhaps $750. 1678pp. 8½ x 12¼. 23147-X Clothbd. **$75.00**

MANUAL OF THE TREES OF NORTH AMERICA, Charles S. Sargent. The basic survey of every native tree and tree-like shrub, 717 species in all. Extremely full descriptions, information on habitat, growth, locales, economics, etc. Necessary to every serious tree lover. Over 100 finding keys. 783 illustrations. Total of 986pp. 5⅜ x 8½. 20277-1, 20278-X Pa., Two-vol. set **$12.00**

GREAT NEWS PHOTOS AND THE STORIES BEHIND THEM, John Faber. Dramatic volume of 140 great news photos, 1855 through 1976, and revealing stories behind them, with both historical and technical information. Hindenburg disaster, shooting of Oswald, nomination of Jimmy Carter, etc. 160pp. 8¼ x 11. 23667-6 Pa. **$6.00**

CRUICKSHANK'S PHOTOGRAPHS OF BIRDS OF AMERICA, Allan D. Cruickshank. Great ornithologist, photographer presents 177 closeups, groupings, panoramas, flightings, etc., of about 150 different birds. Expanded *Wings in the Wilderness.* Introduction by Helen G. Cruickshank. 191pp. 8¼ x 11. 23497-5 Pa. $7.95

AMERICAN WILDLIFE AND PLANTS, A. C. Martin, et al. Describes food habits of more than 1000 species of mammals, birds, fish. Special treatment of important food plants. Over 300 illustrations. 500pp. 5⅜ x 8½. 20793-5 Pa. **$6.50**

THE PEOPLE CALLED SHAKERS, Edward D. Andrews. Lifetime of research, definitive study of Shakers: origins, beliefs, practices, dances, social organization, furniture and crafts, impact on 19th-century USA, present heritage. Indispensable to student of American history, collector. 33 illustrations. 351pp. 5⅜ x 8½. 21081-2 Pa. **$5.50**

OLD NEW YORK IN EARLY PHOTOGRAPHS, Mary Black. New York City as it was in 1853-1901, through 196 wonderful photographs from N.-Y. Historical Society. Great Blizzard, Lincoln's funeral procession, great buildings. 228pp. 9 x 12. 22907-6 Pa. **$9.95**

MR. LINCOLN'S CAMERA MAN: MATHEW BRADY, Roy Meredith. Over 300 Brady photos reproduced directly from original negatives, photos. Jackson, Webster, Grant, Lee, Carnegie, Barnum; Lincoln; Battle Smoke, Death of Rebel Sniper, Atlanta Just After Capture. Lively commentary. 368pp. 8⅜ x 11¼. 23021-X Pa. **$11.95**

TRAVELS OF WILLIAM BARTRAM, William Bartram. From 1773-8, Bartram explored Northern Florida, Georgia, Carolinas, and reported on wild life, plants, Indians, early settlers. Basic account for period, entertaining reading. Edited by Mark Van Doren. 13 illustrations. 141pp. 5⅜ x 8½. 20013-2 Pa. $6.00

THE GENTLEMAN AND CABINET MAKER'S DIRECTOR, Thomas Chippendale. Full reprint, 1762 style book, most influential of all time; chairs, tables, sofas, mirrors, cabinets, etc. 200 plates, plus 24 photographs of surviving pieces. 249pp. 9⅞ x 12¾. 21601-2 Pa. $8.95

AMERICAN CARRIAGES, SLEIGHS, SULKIES AND CARTS, edited by Don H. Berkebile. 168 Victorian illustrations from catalogues, trade journals, fully captioned. Useful for artists. Author is Assoc. Curator, Div. of Transportation of Smithsonian Institution. 168pp. 8½ x 9½. 23328-6 Pa. $6.50

SECOND PIATIGORSKY CUP, edited by Isaac Kashdan. One of the greatest tournament books ever produced in the English language. All 90 games of the 1966 tournament, annotated by players, most annotated by both players. Features Petrosian, Spassky, Fischer, Larsen, six others. 228pp. 5⅜ x 8½. 23572-6 Pa. $3.50

ENCYCLOPEDIA OF CARD TRICKS, revised and edited by Jean Hugard. How to perform over 600 card tricks, devised by the world's greatest magicians: impromptus, spelling tricks, key cards, using special packs, much, much more. Additional chapter on card technique. 66 illustrations. 402pp. 5⅜ x 8½. (Available in U.S. only) 21252-1 Pa. $5.95

MAGIC: STAGE ILLUSIONS, SPECIAL EFFECTS AND TRICK PHOTOGRAPHY, Albert A. Hopkins, Henry R. Evans. One of the great classics; fullest, most authorative explanation of vanishing lady, levitations, scores of other great stage effects. Also small magic, automata, stunts. 446 illustrations. 556pp. 5⅜ x 8½. 23344-8 Pa. $6.95

THE SECRETS OF HOUDINI, J. C. Cannell. Classic study of Houdini's incredible magic, exposing closely-kept professional secrets and revealing, in general terms, the whole art of stage magic. 67 illustrations. 279pp. 5⅜ x 8½. 22913-0 Pa. $5.95

HOFFMANN'S MODERN MAGIC, Professor Hoffmann. One of the best, and best-known, magicians' manuals of the past century. Hundreds of tricks from card tricks and simple sleight of hand to elaborate illusions involving construction of complicated machinery. 332 illustrations. 563pp. 5⅜ x 8½. 23623-4 Pa. $6.95

THOMAS NAST'S CHRISTMAS DRAWINGS, Thomas Nast. Almost all Christmas drawings by creator of image of Santa Claus as we know it, and one of America's foremost illustrators and political cartoonists. 66 illustrations. 3 illustrations in color on covers. 96pp. 8⅜ x 11¼. 23660-9 Pa. $3.50

FRENCH COUNTRY COOKING FOR AMERICANS, Louis Diat. 500 easy-to-make, authentic provincial recipes compiled by former head chef at New York's Fitz-Carlton Hotel: onion soup, lamb stew, potato pie, more. 309pp. 5⅜ x 8½. 23665-X Pa. $3.95

SAUCES, FRENCH AND FAMOUS, Louis Diat. Complete book gives over 200 specific recipes: bechamel, Bordelaise, hollandaise, Cumberland, apricot, etc. Author was one of this century's finest chefs, originator of vichyssoise and many other dishes. Index. 156pp. 5⅜ x 8. 23663-3 Pa. $2.95

TOLL HOUSE TRIED AND TRUE RECIPES, Ruth Graves Wakefield. Authentic recipes from the famous Mass. restaurant: popovers, veal and ham loaf, Toll House baked beans, chocolate cake crumb pudding, much more. Many helpful hints. Nearly 700 recipes. Index. 376pp. 5⅜ x 8½. 23560-2 Pa. $4.95

ILLUSTRATED GUIDE TO SHAKER FURNITURE, Robert Meader. Director, Shaker Museum, Old Chatham, presents up-to-date coverage of all furniture and appurtenances, with much on local styles not available elsewhere. 235 photos. 146pp. 9 x 12. 22819-3 Pa. $6.95

COOKING WITH BEER, Carole Fahy. Beer has as superb an effect on food as wine, and at fraction of cost. Over 250 recipes for appetizers, soups, main dishes, desserts, breads, etc. Index. 144pp. 5⅜ x 8½. (Available in U.S. only) 23661-7 Pa. $3.00

STEWS AND RAGOUTS, Kay Shaw Nelson. This international cookbook offers wide range of 108 recipes perfect for everyday, special occasions, meals-in-themselves, main dishes. Economical, nutritious, easy-to-prepare: goulash, Irish stew, boeuf bourguignon, etc. Index. 134pp. 5⅜ x 8½. 23662-5 Pa. $3.95

DELICIOUS MAIN COURSE DISHES, Marian Tracy. Main courses are the most important part of any meal. These 200 nutritious, economical recipes from around the world make every meal a delight. "I . . . have found it so useful in my own household,"—*N.Y. Times*. Index. 219pp. 5⅜ x 8½. 23664-1 Pa. $3.95

FIVE ACRES AND INDEPENDENCE, Maurice G. Kains. Great back-to-the-land classic explains basics of self-sufficient farming: economics, plants, crops, animals, orchards, soils, land selection, host of other necessary things. Do not confuse with skimpy faddist literature; Kains was one of America's greatest agriculturalists. 95 illustrations. 397pp. 5⅜ x 8½. 20974-1 Pa. $4.95

A PRACTICAL GUIDE FOR THE BEGINNING FARMER, Herbert Jacobs. Basic, extremely useful first book for anyone thinking about moving to the country and starting a farm. Simpler than Kains, with greater emphasis on country living in general. 246pp. 5⅜ x 8½. 23675-7 Pa. $3.95

PAPERMAKING, Dard Hunter. Definitive book on the subject by the foremost authority in the field. Chapters dealing with every aspect of history of craft in every part of the world. Over 320 illustrations. 2nd, revised and enlarged (1947) edition. 672pp. 5⅜ x 8½. 23619-6 Pa. $8.95

THE ART DECO STYLE, edited by Theodore Menten. Furniture, jewelry, metalwork, ceramics, fabrics, lighting fixtures, interior decors, exteriors, graphics from pure French sources. Best sampling around. Over 400 photographs. 183pp. 8⅜ x 11¼. 22824-X Pa. $6.95

ACKERMANN'S COSTUME PLATES, Rudolph Ackermann. Selection of 96 plates from the *Repository of Arts*, best published source of costume for English fashion during the early 19th century. 12 plates also in color. Captions, glossary and introduction by editor Stella Blum. Total of 120pp. 8⅜ x 11¼. 23690-0 Pa. $5.00

THE ANATOMY OF THE HORSE, George Stubbs. Often considered the great masterpiece of animal anatomy. Full reproduction of 1766 edition, plus prospectus; original text and modernized text. 36 plates. Introduction by Eleanor Garvey. 121pp. 11 x 14¾. 23402-9 Pa. **$8.95**

BRIDGMAN'S LIFE DRAWING, George B. Bridgman. More than 500 illustrative drawings and text teach you to abstract the body into its major masses, use light and shade, proportion; as well as specific areas of anatomy, of which Bridgman is master. 192pp. 6½ x 9¼. (Available in U.S. only) 22710-3 Pa. **$4.50**

ART NOUVEAU DESIGNS IN COLOR, Alphonse Mucha, Maurice Verneuil, Georges Auriol. Full-color reproduction of *Combinaisons ornementales* (c. 1900) by Art Nouveau masters. Floral, animal, geometric, interlacings, swashes—borders, frames, spots—all incredibly beautiful. 60 plates, hundreds of designs. 9⅜ x 8-1/16. 22885-1 Pa. **$4.50**

FULL-COLOR FLORAL DESIGNS IN THE ART NOUVEAU STYLE, E. A. Seguy. 166 motifs, on 40 plates, from *Les fleurs et leurs applications decoratives* (1902): borders, circular designs, repeats, allovers, "spots." All in authentic Art Nouveau colors. 48pp. 9⅜ x 12¼. 23439-8 Pa. **$6.00**

A DIDEROT PICTORIAL ENCYCLOPEDIA OF TRADES AND INDUSTRY, edited by Charles C. Gillispie. 485 most interesting plates from the great French Encyclopedia of the 18th century show hundreds of working figures, artifacts, process, land and cityscapes; glassmaking, papermaking, metal extraction, construction, weaving, making furniture, clothing, wigs, dozens of other activities. Plates fully explained. 920pp. 9 x 12. 22284-5, 22285-3 Clothbd., Two-vol. set **$50.00**

HANDBOOK OF EARLY ADVERTISING ART, Clarence P. Hornung. Largest collection of copyright-free early and antique advertising art ever compiled. Over 6,000 illustrations, from Franklin's time to the 1890's for special effects, novelty. Valuable source, almost inexhaustible.
Pictorial Volume. Agriculture, the zodiac, animals, autos, birds, Christmas, fire engines, flowers, trees, musical instruments, ships, games and sports, much more. Arranged by subject matter and use. 237 plates. 288pp. 9 x 12. 20122-8 Clothbd. $15.95

Typographical Volume. Roman and Gothic faces ranging from 10 point to 300 point, "Barnum," German and Old English faces, script, logotypes, scrolls and flourishes, 1115 ornamental initials, 67 complete alphabets, more. 310 plates. 320pp. 9 x 12. 20123-6 Clothbd. **$16.95**

CALLIGRAPHY (CALLIGRAPHIA LATINA), J. G. Schwandner. High point of 18th-century ornamental calligraphy. Very ornate initials, scrolls, borders, cherubs, birds, lettered examples. 172pp. 9 x 13. 20475-8 Pa. **$7.95**